Mastering

Human biology

P-

599.0

-3 MR
31 MAR 1993
25 OCT 1993
13 DEC 1993

11 APR 1994
-9 MAY 1994

-8 JUN 1994

29 JUN 1994

17 OCT 1994

-2 MAY 1995

Macmillan Master Series

Mastering

Human biology

Jean Roberts

MACMILLAN

First published 1991

10 9 8 7 6 5 4 3 2
00 99 98 97 96 95 94 93 92 91

Published by
MACMILLAN EDUCATION LTD
Houndmills, Basingstoke, Hampshire RG21 2XS
and London
Companies and representatives
throughout the world

Typeset and Ilustrated by TecSet Ltd, Wallington, Surrey
Printed in Hong Kong

British Library Cataloguing in Publication Data
Roberts, Jean
Mastering human biology
1. Man. Physiology
I. Title
612
ISBN 0 – 333 – 52042 – 4 Pbk
ISBN 0 – 333 – 52043 – 2 Pbk export

For my parents, Anne and Frank

Contents

Preface

The decision to study human biology often stems from an interest in how the body works. Enthusiastic students may, however, feel alienated and inadequate when they encounter mathematics and chemistry within the course, and irritated by the appearance of plants and non-human animals – these may seem out of place and unnecessary. This book is designed to enable the student to build up a good understanding and knowledge of the human being, not only as a working body but also as one that interacts with its planet and its fellow-occupants. The student will also realise that some chemistry and mathematics are integral to developing this understanding. Study of the subject crosses artificial academic barriers and enables the growth of scientific skills.

The aims of GCSE human biology courses have fortunately moved us away from a cold, static approach and towards real science, real understanding and real enjoyment. *Mastering Human Biology* fulfils all the aims and objectives of GCSE and its subject content will meet the requirements of all the GCSE boards.

This book is suitable for any student of human biology, whether he or she is following a full-time or a part-time course. Open learning students will find it easy to follow; the ideas and themes build upon each other progressively. Lack of previous scientific knowledge will not be a handicap, as new concepts are explained at the time the student needs to know them.

Parts of the text are boxed, either for easy reference or to provide the student with information that may be unfamiliar, such as the commonly used units of measurement; students who do possess this knowledge can disregard that particular box, while others will have easy access to the information they need without having to find other textbooks and thereby lose the flow of the text.

The book is designed to be interactive – that is, not just to provide 'facts' but to enable students to think and develop their own ideas. For example, students are encouraged to predict and deduce, and to design experiments, all of which will enable them to 'own' the subject. *Mastering Human Biology* will act as a facilitator, encouraging curiosity, appreciation, interest and enjoyment as well as developing scientific method and skills.

JEAN ROBERTS

A message to the student

How to use this book

This book is designed to help you understand and learn, as well as being a source of information.

You will find exercises for you to do at intervals throughout the text. Doing these exercises will help you to understand and use what you have been studying.

Some notes are boxed. These notes are either for quick reference or to explain a topic or an idea used in the text with which you might not be familiar.

As the text progresses there are memory checks with section references. They will remind you of ideas you have already come across earlier in the book.

At the end of each section there is a summary and questions. The questions will help you use the material you have been studying and give you practice in expressing your thoughts.

Pay special attention to the instructions for the experiments. *Always* wash your hands carefully before and after handling plant or animal material or doing microbiological work. Cover cuts and scratches with a waterproof dressing.

Experiments marked with this symbol:

Should only be done *in a laboratory under qualified supervision. Never* attempt to do these by yourself.

Acknowledgements

The author and publishers wish to thank the following who have kindly given permission for the use of copyright material: Don Mackean for Figs 1.8, 9.1 and 11.15(a) and (b) from *Introduction to Biology* by Don G. Mackean, John Murray.

The author and publishers wish to acknowledge the following illustration suppliers: Barnaby's Picture Library; Biofotos Associates; Bioset More Education; Bubbles Picture Library; Cambridge Scientific Instruments; Camera Press Ltd; Clark's Shoes Ltd; Colorsport; Format; Philip Harris (Medical) Ltd; Macmillan Archive; Metropolitan Police; Jean Roberts; Science Photo Library; Harry Smith Horticultural Photographic Collection; Alan Thomas.

Every effort has been made to trace all the copyright holders but if any have been inadvertently overlooked the publishers will be pleased to make the necessary arrangement at the first opportunity.

The author also wishes to thank Dr Angela Roberts for her advice and expertise, and Dr Jean Macqueen for her continued support.

1 **Building blocks**

The study of human biology will help you to know and understand how your body is constructed and how it works. Human beings do not live in isolation, however. They are dependent upon each other and upon other living things.

This chapter and the two following chapters will provide a foundation for your study. We will examine some processes that are fundamental to the working of the human body, and how these processes are inter-related.

1.1 **The basic unit of life**

Living things are made up of box-like compartments called *cells*. Some living things consist of just one cell; others are made up of millions of cells, all working together.

Exercise

Fig. 1.1 shows drawings of three different cells. Study them for a few minutes.

Look for and note down similarities and differences between them. (You do not need to name the parts of the cell for this exercise.)

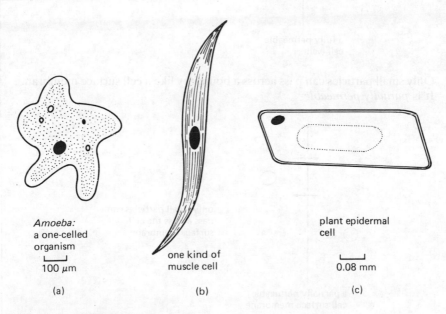

Amoeba:
a one-celled
organism
⊢——⊣
100 µm

(a)

one kind of
muscle cell

(b)

plant epidermal
cell

⊢——⊣
0.08 mm

(c)

Figure 1.1   Different kinds of cell

We are going to concentrate on the similarities shared by the cells in Fig. 1.1. (Refer to the notes you made in the exercise as you go through the descriptions that follow.)

The black area within all three cells is the *nucleus*. It controls the chemical reactions occurring in the cell. It is, therefore, a store of information. This information must be passed on to new cells, and you will see how this happens when you study cell division. The information in the nucleus exists as a code, arranged on long thin threads called *chromosomes*. Only a few types of human cell do not have a nucleus.

The outer boundary of a plant cell (Fig. 1.1(c)) is a rigid fully permeable wall. (To help you understand what this means, read the boxed note 'Permeable and partially permeable'.) The wall is made of *cellulose*. When we eat plants like apples or carrots, cellulose gives us 'fibre' or 'roughage'. The wall gives shape to the cell, and ultimately to the plant. It is called the *cell wall*.

Permeable and partially permeable

Quite large particles, as well as small ones, can pass across a boundary like a cell wall. We say it is *fully permeable*.

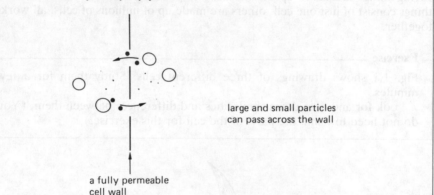

large and small particles
can pass across the wall

a fully permeable
cell wall

Only small particles can pass across a boundary like a cell surface membrane. It is *partially permeable*.

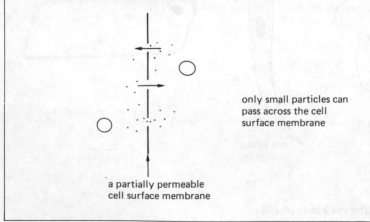

only small particles can
pass across the cell
surface membrane

a partially permeable
cell surface membrane

The outer boundary of animal cells (Figs. 1.1(a) and (b)) is the *cell surface membrane*. It is flexible and partially permeable. Plant cells have a cell surface membrane too. It lies just inside the cell wall.

The material surrounding the nucleus and enclosed by the cell surface membrane is *cytoplasm*. It has a jelly-like consistency and is the site of many chemical reactions. The cytoplasm contains very small structures called *organelles*, which are only visible under high-powered microscopes. Certain chemical reactions occur only in or on these organelles.

A plant cell and an animal cell

We are going to compare two simple cells: one from the human body, and one from a plant.

The human cell (look at Fig. 1.2) is from the inside of the cheek. Cells like this are simple flat covering cells, called *pavement cells* or *squamous epithelium*.

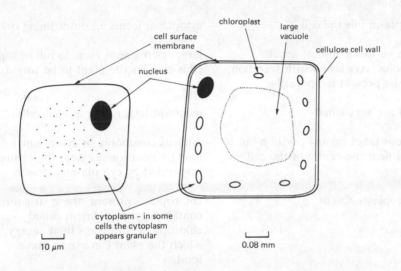

cell surface membrane	chloroplast
nucleus	large vacuole
	cellulose cell wall
cytoplasm – in some cells the cytoplasm appears granular	

10 μm

0.08 mm

Figure 1.2 Cheek cell   **Figure 1.3 Plant mesophyll cell**

Fig. 1.3 is a drawing of a plant cell. Cells like this are found in the lower layers of a leaf. They are loosely packed together, forming a spongy layer. They are called *mesophyll cells*.

Exercise

Look at the cheek cell in Fig. 1.2. Find

the cell surface membrane
the nucleus
the cytoplasm

Now examine both Figs. 1.2 and 1.3.

Make a note of the similarities and differences between the cheek cell and the mesophyll cell.

Then compare your notes with those in Table 1.1. The table mentions some differences which are not obvious from the drawings.

Table 1.1   A comparison of a plant and an animal cell

Animal	*Plant*
cell surface membrane as outer boundary, which allows flexibility (it allows the cell to move around and the likelihood of being damaged is reduced)	cellulose wall as outer boundary; cell surface membrane lies on the inside of the cell wall
cytoplasm fills the cell	cytoplasm forms an outer lining only
large vacuoles absent; small vacuoles, concerned with secretion, may be present temporarily	large, permanent vacuole full of sap (this enables the plant to be turgid)
most are very small	most are larger than animal cells
nucleus takes up any position but is often near the centre of the cell	nucleus commonly in cytoplasm lining – occasionally near the centre, suspended by cytoplasmic strands
chloroplasts absent	chloroplasts present: these structures contain a green pigment called chlorophyll; this traps light energy, which the plant can use to make food
carbohydrate, an energy-giving food, stored as glycogen	carbohydrate, an energy-giving food, stored as starch

Summary

■ Living things are made up of cells.

■ An animal cell has a cell surface membrane, cytoplasm and a nucleus.

■ A plant cell has a cell wall, a cell surface membrane, cytoplasm, a nucleus, a vacuole and (if the cell is green) chloroplasts.

Look at Fig. 1.4. It shows two cells, labelled A and B. One is a plant cell and one is an animal cell.

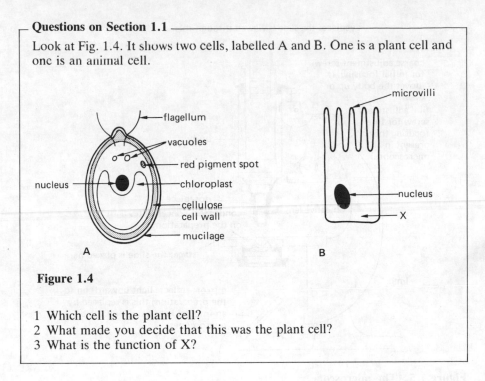

Figure 1.4

1 Which cell is the plant cell?
2 What made you decide that this was the plant cell?
3 What is the function of X?

1.2 Observing cells

Most cells are very small, generally around 100 μm across. (If you are not sure what that means read the boxed note 'Units of measurement'.) So we need to use microscopes to study cells. There are two main types: the light microscope and the electron microscope.

Units of measurement

1 μm = 1/1000 mm	μm = micrometre
1000 μm = 1 mm	mm = millimetre
10 mm = 1 cm	cm = centimetre
100 cm = 1 m	m = metre

A *light microscope* is used for examining cell preparations requiring a magnification up to ×900. An *electron microscope* magnifies up to ×500 000 and is used to examine very small structures in cells.

Look at Fig. 1.5. This is a diagram of a light microscope. Study it carefully. Find the *eyepiece*. This is what you look through.

Find the *objectives*. Many light microscopes have three objectives. Each objective has a lens.

The power of magnification of a lens is usually marked on its edge. Most eyepiece lenses are ×10, ×15 or ×20. Most objective lenses are ×4, ×10, ×20 or

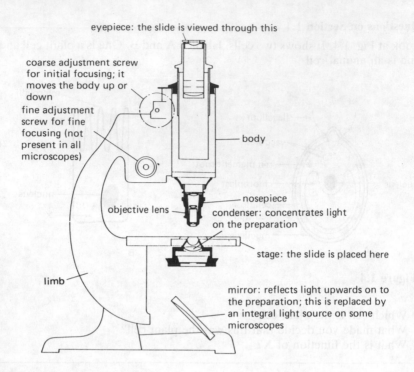

coarse adjustment screw for initial focusing; it moves the body up or down

fine adjustment screw for fine focusing (not present in all microscopes)

eyepiece: the slide is viewed through this

body

nosepiece

objective lens

condenser: concentrates light on the preparation

stage: the slide is placed here

limb

mirror: reflects light upwards on to the preparation; this is replaced by an integral light source on some microscopes

Figure 1.5   The microscope

×40. If you want to use ×400 magnification, for example, use a ×10 eyepiece and a ×40 objective (10 × 40 = 400).

You have to position the objective you want to use under the eyepiece and tube, by rotating the nosepiece.

The microscope slide is placed on the *stage*. Find the stage in the drawing. The slide can be held in place by metal clips on the stage.

Find the *mirror*. Light is reflected upwards from the mirror. Light must pass up through the hole in the stage, through the slide, through the objective and then through the eyepiece.

Find the coarse and fine *adjustment screws*. Movement of these screws will allow you to view the slide clearly.

Use of the microscope

1 Look down the eyepiece. Move the mirror until the field of view is bright. If the light source is integral (if there is a lamp instead of a mirror) switch it on. (**Never** use sunlight as your light source. It can damage your eyes. Use a lamp.)
2 Move the rotating nosepiece so that the lowest-power objective is positioned over the hole in the stage.
3 Put the slide on the stage. Secure it with the clips. Make sure the preparation you want to view is placed centrally over the light source.
4 Using the coarse adjustment, move the objective as close to the slide as possible.

5 Look down the eyepiece. Move the coarse adjustment slowly, until the preparation becomes focused.
6 Use the fine adjustment to bring the preparation into sharper focus.
7 Scan the slide to find an area to be examined.
8 Rotate the nosepiece, to use a higher-power objective.
9 Use the fine adjustment for final focusing.

Always clean the lenses with lens tissue soaked in propanol.

Wipe up any liquid spilled on the microscope immediately. Only use fine adjustment for high-power objectives. (If your microscope only has a coarse adjustment, adjust the objective with care to avoid breaking the slide.)

Never pull the objectives. Use the nosepiece to change their position.

After examining the preparation, remove the slide and put the low-power objective in position.

Making a cell preparation

Specimens are mounted on glass microscope slides, usually in a drop of liquid. They are often stained to make them easier to examine. The preparation is held flat, using a small piece of glass called a *coverslip*.

A suitable specimen for observation is onion epidermis. (This is an outer layer of cells, like a skin.) You should be able to see the nucleus, cytoplasm and cell walls of the epidermal cells.

You will need:

clean microscope slide
clean coverslip
scalpel
forceps

probe or mounted needle
iodine solution
onion

Method
1 Cut a ½ cm square of the white flesh of the onion.
2 Gently scratch the edge of the inside surface with a scalpel until a transparent skin lifts up.
3 Pull this skin off with the forceps. It should come away easily. This is a single layer of cells.
4 Place the skin centrally on the slide.
5 Add two or three drops of iodine solution. This will stain the nucleus orange and make it visible.
6 Place the coverslip over the onion skin, trying to minimise air bubbles. Start by putting the coverslip on the edge of the iodine drop. Support it with a mounted needle or probe.

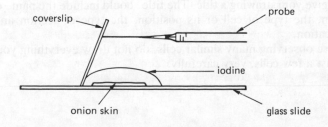

Gently lower the coverslip down over the onion skin.

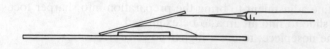

Finally, remove the probe.

7 Put the slide on the stage of the microscope and examine it as described in the last section.
8 Draw one or two cells using a sharp pencil. (When drawing biological specimens or apparatus you should follow the guidelines set out in the boxed note 'Drawings'.)

Look at Fig. 1.6(a), which is a drawing of an onion epidermal cell. Fig. 1.6(b) is a photograph of onion epidermal cells as seen under a microscope. The cell surface membrane cannot usually be seen with a light microscope. This is partly because it is pressed against the cell wall, and partly because it is very thin.

Drawings ─────────────────────────────────────

Here are some guidelines to follow when you are making biological drawings.

1 Always use a sharp pencil. An HB pencil is suitable.
2 Make a faint outline to begin with. When completing the drawing keep your pencil on the paper until a particular line is completed.
3 Make your drawings *large*; even a drawing of just one cell should be at least 5 cm across. Larger than this is better as it allows structures to be seen easily.
4 Never use a ruler when you are drawing cells and living materials.
5 Use a ruler for glassware where appropriate.
6 Do not shade your drawings. Clean simple lines are all that is required.
7 Use a ruler for label lines. They should not cross each other and they must be carefully positioned.
8 Print labels in pencil.
9 Always give your drawing a title. The title should include the name of the organism, the type of cell or its position, the type of section and the magnification.
10 If you are observing many similar cells, do not draw everything you see. Draw just a few cells, very carefully.

(a)

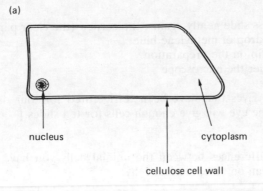

(b)

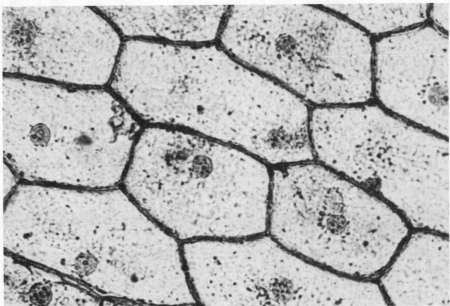

Figure 1.6 Onion epidermal cells seen under the light microscope: (a) a drawing, (b) a photograph

EXPERIMENT

Aim: To observe animal cells with a microscope

You will need:
a fresh bullock's eye
glass slide
coverslip

mounted needle
microscope
methylene blue stain

Method

1 Press the glass slide gently against the cornea (the clear part) of the eye.
2 Stain with a drop of methylene blue.
3 Put a coverslip on the preparation.
4 Examine under the microscope.

[*Teacher's note*: Eyes that have been refrigerated for four days will still give corneal cells. One eye will give enough cells for ten slides.]

Question

What are the differences between the animal cells you have observed in this experiment and an onion epidermal cell?

Questions on Section 1.2

1 The eyepiece on a microscope is marked '× 10'. What magnification is obtained if the following objectives are used?
(a) ×4   (b) ×10   (c) ×20

2 A student wishes to examine a slide of blood cells with the instrument shown in Fig. 1.5.
(a) Explain how she could set the instrument up for use.
(b) Explain how she could view the slide under high magnification.

3 (a) Explain how you could prepare a microscope slide ready for viewing plant tissue under a microscope.
(b) Give the name of a suitable plant tissue that you could use.
(c) What apparatus would you need for this preparation?

1.3 **Variety, design and function**

In this section we will look at some of the different types of human cell, and examine how their structure enables them to fulfil particular functions.

Since the cells we shall consider are human, they are animal cells. The features you should expect to see in each cell are the nucleus, the cytoplasm and the cell surface membrane.

The cheek cell you studied in Section 1.1 can be described as 'simple'. It is a covering or lining cell; it is flat and many such cells together form a lining to the mouth. A layer of cells like this is called an *epithelium*.

Now look at Fig. 1.7. This shows a type of nerve cell. Nerve cells are called *neurones*. The figure shows a *receptor neurone* and it is highly specialised – that means it has special characteristics that enable it to fulfil its particular function. Receptor neurones are responsible for picking up messages from the skin, eye, ear, tongue or nose (the sense organs) and taking these messages in the form of nervous impulses to the spinal column or the brain.

When studying any new cell, first find the nucleus, cytoplasm and cell surface membrane. All remaining structures are there for a reason. Cells are usually well designed! To understand the design of a cell, we must ask ourselves two questions:

'What does this cell do?'

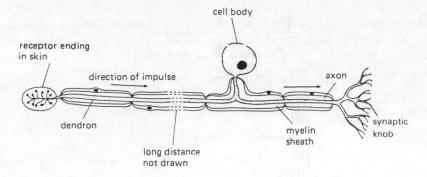

Figure 1.7 A receptor neurone

and then,

'How is it designed to fulfil this function?'

In relation to the receptor neurone, the answer to the first question would be:

- the cell must be able to pick up information
- it must be able to pass it on (like electricity)
- it must be able to deliver the message
- it must do all of these things quickly.

Now the second question:

- the receptor neurone endings *pick up* information
- the dendron and axon *pass it on*
- the synaptic knobs *deliver it* to another neurone
- the dendron fibre is long. The dendron of a receptor neurone ending on the tip of your big toe will stretch from your toe to your spinal column – this is a long way for a cell! (Cell bodies are always found just outside the spinal column, so the axon is relatively short.) One long dendron means the message can be *passed on quickly*.

Look at Fig. 1.7 again. Find the receptor neurone endings, the dendron, the axon and the synaptic knobs. Try to remember what they do. Note that the cell body lies to one side of the fibre. It does not interfere with the movement of the message. (A signal box is not in the middle of a railway track, it stands at one side.) The fibre branches at the end. (Find the branching on the diagram.) Each branch has a tiny swelling called a synaptic knob. You will learn in Section 10.2 that synaptic knobs pass the message on to other nervous tissue. The branching ensures that the message is passed on quickly. It also means that it can be passed on to more than one cell. This allows for many responses to be made.

We have spent a long time studying this one cell, but this should help you to look at other cells much more critically.

Let's look at another neurone. Look at Fig. 1.8. This is a effector neurone. It is responsible for taking information from the brain or spinal cord and passing it on to muscles, such as those in a limb. The muscles contract and the limb moves. Look carefully at the drawing, and find the nucleus, the cytoplasm and the cell surface membrane. The other structures you can see all have a purpose.

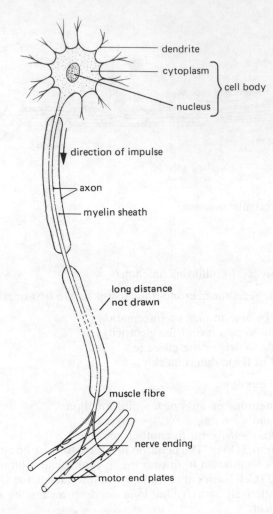

Figure 1.8 An effector neurone

Exercise

Write down how you think the dendrites, axon and motor end plates enable the cell to fulfil its function. Compare your responses to those in the table at the end of the chapter, headed 'Effector neurone analysis'.

When you meet a new cell, you can practise these analytical skills. This will enable you to understand what the cell does and to remember its structure.

We cannot study every type of cell here, but there are a few unusual cells that deserve special mention. Look carefully at Figs. 1.9, 1.10 and 1.11. Notice that all these drawings have a scale. This will enable you to work out how big the cells are. (Read the boxed note 'How big are cells?')

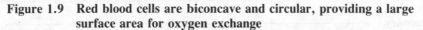

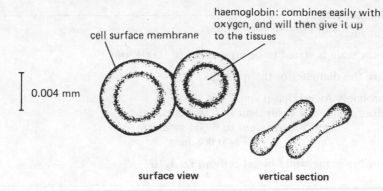

cell surface membrane

haemoglobin: combines easily with oxygen, and will then give it up to the tissues

0.004 mm

surface view

vertical section

Figure 1.9 **Red blood cells are biconcave and circular, providing a large surface area for oxygen exchange**

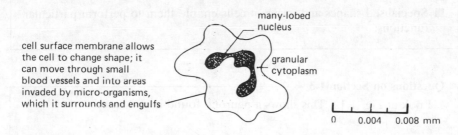

cell surface membrane allows the cell to change shape; it can move through small blood vessels and into areas invaded by micro-organisms, which it surrounds and engulfs

many-lobed nucleus

granular cytoplasm

0   0.004   0.008 mm

Figure 1.10 **White blood cell (phagocyte)**

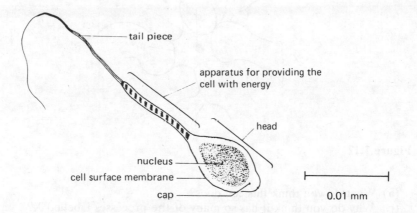

tail piece

apparatus for providing the cell with energy

head

nucleus

cell surface membrane

cap

0.01 mm

Figure 1.11 **Sperm cell. Notice how little cytoplasm it has. This cell is designed to carry nuclear material into the female reproductive tract and fuse with the female sex cell. It will not function as an independent cell for more than a few days, so little cytoplasm is required**

How big are cells? ────────────────────────────

Look at Fig. 1.9.

The 1 cm scale is stated to be equivalent to 0.004 mm.

Measure the diameter of the red cell.

You probably found that it was 2 cm.
Therefore, if 1 cm is equivalent to 0.004 mm,
            2 cm is equivalent to 0.004 × 2 mm
                         = 0.008 mm

So, how big is the white blood cell in Fig. 1.10?

Summary ────────────────────────────────────

■ All cells have the same basic structure.

■ Specialised shapes and forms of cells enable them to perform particular
   functions.

Questions on Section 1.3 ───────────────────────

1 Look at Fig. 1.12. This shows a neurone found in the brain.

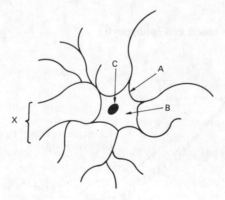

Figure 1.12

  (a) What do you think this cell does?
  (b) Why do you think it has so many of the processes labelled X?
  (c) What are the structures labelled A, B and C?

2 Look at Fig. 1.13.
  (a) What is the size of this cell across XY?
  (b) Draw a diagram to show the appearance of this cell if it was sectioned
      across XY.

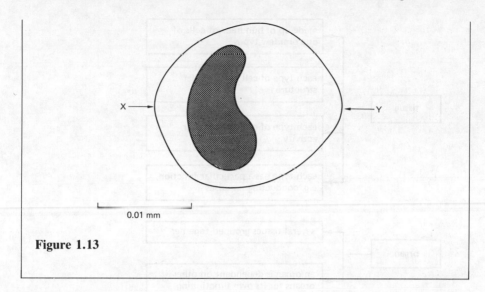

0.01 mm

Figure 1.13

1.4 Cells working together

From previous sections we can say that

- our bodies are composed of cells
- the cells have different shapes, sizes, structures and properties. A cell's special characteristics enable it to fulfil its particular function. That is, cells are *specialised*.

Human cells do not exist in isolation. They cannot live and work without similar cells around them. A muscle in your arm is made up of thousands of muscle cells. Your brain is made up of thousands of neurones. Many similar cells performing similar functions form a *tissue*. Several tissues grouped together make up a working unit called an *organ*. For example, the heart is an organ made up of muscle tissue, lining tissue, nerve tissue and tissue forming an enclosing sac. A series of organs whose functions are co-ordinated form a *system*, such as the digestive system or the circulatory system. Systems functioning together make up an *organism*. Fig. 1.14 is a summary of the characteristics of tissues, organs and systems.

── Summary ────────────────────────────

■ The human body is composed of cells, tissues, organs and systems.

■ These systems work together to form the organism.

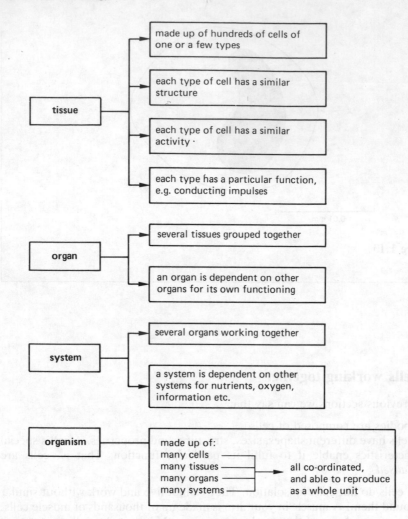

Figure 1.14   Tissues, organs, systems and organisms

Question on Section 1.4

Which of the following are cells, which are tissues and which are organs?

blood	effector neurone
kidney	brain
heart	skin
muscle	heart muscle (cardiac muscle)
sperm	liver

1.5 **Characteristics of living things**

Respiration

This takes place in every cell. To respire, a cell usually needs oxygen. We breathe in oxygen, which is then transported to each cell in the body by the bloodstream.

In respiration the energy in glucose (food) is released. Glucose is broken down into smaller particles. These particles combine with oxygen to form carbon dioxide and water. This can be summarised in a sum or equation:

food + oxygen = carbon dioxide + water + energy

We respire so that we can use the energy stored in food to do work. Some examples of the work the human body does are: muscle contraction, repairing and maintaining tissues, growth and chemical reactions in the cell.

Plants respire in the same way. They too use oxygen to release the energy from food.

Feeding

We need to eat so that we

- have energy for work
- have materials for growth and repair of tissues.

Foods are divided up into groups depending on their structure and what they do for us. These groups are called the *nutrients*. A summary of the nutrients is given in Table 1.2. You will find out more about them later in this book.

Excretion

Excretory products are made by the body as a result of chemical processes in its cells. If they are not eliminated they start to build up in the body. Too great a build-up is fatal. That is, they act as poisons. Here is a summary of where excretory products come from and how the body gets rid of them.

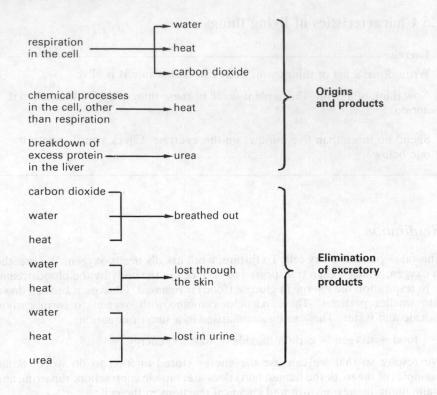

Table 1.2 Nutrients and their functions

Nutrient	Function
protein	growth and repair cell surface membranes and enzymes
carbohydrates	production of energy
lipids	production of energy insulation cell surface membranes
vitamins	chemical reactions in cells
minerals	chemical reactions in cells making bones and teeth
water	all chemical reactions in the body take place in water
roughage (fibre)	allows quick disposal of undigested food for a healthy digestive system

Plants pass excretory products that are gases out of the leaves. They store other excretory products: for example, in leaves that will be shed and as oils in older leaves.

Growth

There are three aspects of growth:

- organisms get bigger
- cells may increase in number
- cells may become specialised (into muscle or nerve cells, for example); they have become *differentiated*.

Movement

Human beings have a system of muscles connected to bones, which allows us to make movements.

Plants move towards light, gravity and water. The movements are very slow and are caused by unequal growth in stems, leaves or roots. Some plants have special mechanisms allowing them to move quickly; examples are some insect-catching plants, like the sundew, and the sensitive plant *Mimosa pudica*.

Look at Fig. 1.15, which shows two different kinds of movement.

Reproduction

Any organism must be able to produce other organisms like itself. The population of that organism can then continue. Bacteria reproduce themselves by dividing into two. In good conditions they can do this every 20 minutes. In human beings it takes longer, and it is a lot more complicated! A new human being is created when a sex cell from a male combines with a sex cell from a female. This is called *sexual* reproduction.

Plants produce seeds as a result of sexual reproduction. Some plants reproduce *asexually*; for example, a strawberry produces small plantlets at the end of runners.

Look at Fig. 1.16, which shows several organisms that have reproduced.

Sensitivity

A living organism must be able to respond, or be *sensitive*, to changes in its environment. Human beings use their five senses to keep a continuous check on surroundings. This is how we can avoid danger, obtain food, react socially, and so on.

Some plants are sensitive to touch, particularly climbing plants and those like the sundew in Fig. 1.15(b). Plants are sensitive to gravity, light and water.

Fig. 1.17 shows two examples of sensitivity in organisms.

(a)   (b)

Figure 1.15   Movement: (a) footballers have to move quickly; (b) sundews catch insects

(a)

Figure 1.16   (a) Humans and (b) beans reproduce sexually; (c) strawberry runners: an example of asexual reproduction

(b)

(c)

(a)

(b)

Figure 1.17   (a) We act upon information from our senses; (b) some plants are sensitive to touch and respond by attaching themselves to the source of the stimulus

Summary

All living things *respire*, *excrete*, *grow*, *move*, *feed*, *reproduce* and are *sensitive*.

Question for Section 1.5

A bean plant and a rabbit are both living things. Compare the ways in which both feed, move and reproduce. Draw up a table to help present your comparison.

Response to exercise

Effector neurone analysis (see Section 1.3)

Structure	Notes
dendrites	very branched; many impulses can be picked up at one time, perhaps from many other neurones
axon	one axon; the impulse can pass quickly and without interruption to the muscle
motor end plates	branched; since they end in muscle, this enables the impulse to be passed to many muscle cells

2 Cells at work

2.1 Movement in and out of cells

Single-celled organisms and cells that live together all need to get food and oxygen into themselves. And they must get waste out. All of these substances have to pass across the cell surface membrane. We know from Section 1.1 that the membrane is partially permeable. It behaves like a very fine sieve. Small particles can get through it, but large particles cannot. Partial permeability means that there is some control over which particles get in and which particles get out of the cell. There are some questions that must be asked, however.

- The cell needs oxygen to live. What makes oxygen get into the cell?
- How does food get into the cell? Does it escape?
- What makes waste like carbon dioxide pass out of the cell?

To answer these questions we must start by looking at how particles move. Let's consider three situations:

- The smell of paint moves throughout a house, even if the doors are closed and there 'appears' to be no movement of air.
- Look at Fig. 2.1. A drop of dye is placed in a beaker of water, which is left quite still. After a while the dye has coloured the entire beakerful of water.
- A drop of dye is placed in a cavity made in a jelly. Look at Fig. 2.2. After a while the jelly has taken up the colour of the dye.

Situations like these may not surprise you. They happen all the time. But why do they happen?

Liquids and gases are made up of particles that are moving all the time. *They tend to move from an area of high concentration to an area of low concentration. The result is equilibrium or an equal concentration.* Although the particles are still

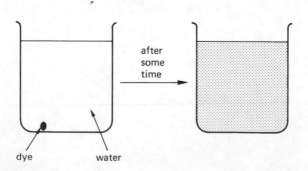

Figure 2.1   Movement of dye in water

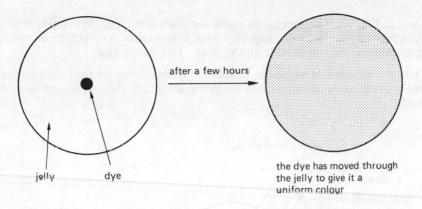

after a few hours

jelly   dye

the dye has moved through
the jelly to give it a
uniform colour

Figure 2.2   Movement of dye in a gel

moving, there appears to be no overall change. The movement of gases and liquids leading to an equal concentration is called *diffusion*.

Diffusion is an important process for living things. For example, the *Amoeba* is a one-celled organism. It gets oxygen across its cell surface membrane by diffusion. Look at Fig. 2.3; notice that the concentration of oxygen in the cell is low. Oxygen particles will move from an area of high concentration to an area of low concentration. Oxygen is being used by the cell all the time, so the concentration will always be lower inside the cell compared to the outside. This results in a concentration gradient.

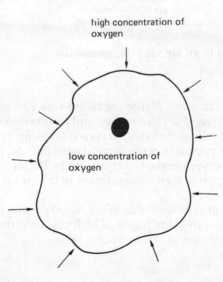

high concentration of
oxygen

low concentration of
oxygen

Figure 2.3   *Amoeba*: oxygen will move across the cell surface and into the organism. As oxygen is being used up all the time, a concentration gradient always exists. Carbon dioxide leaves the *Amoeba* in the same way

Look at Fig. 2.4. This represents an air sac in a human lung. When you breathe in, you take oxygen into your lungs and into these air sacs. You breathe in oxygen and breathe out carbon dioxide all the time. This means that:

- the concentration of oxygen in the air sac is always higher than in the body
- the concentration of carbon dioxide in the air sacs is always lower than in the body.

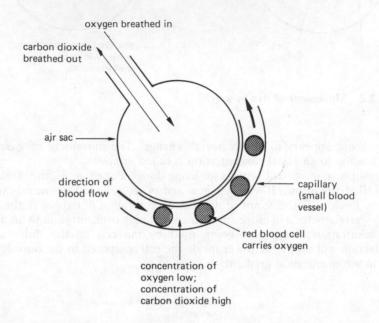

Figure 2.4   Diffusion in an air sac (diagrammatic)

Find the air sac in the drawing. Notice the blood vessel around the air sac. Blood is responsible for carrying gases like oxygen and carbon dioxide around the body. As blood reaches the cells of the body, oxygen is taken up by the cells and carbon dioxide, produced as a result of respiration, passes into the the blood. Therefore the concentration of oxygen in the blood passing by the air sacs is *low*. So oxygen will pass from the region of high concentration in the air sac to the region of low concentration in the blood.

The concentration of carbon dioxide in the blood passing the air sac is *high*. So carbon dioxide will pass from the region of high concentration in the blood to the region of low concentration in the air sac.

Exercise

Diffusion occurs across the lung membranes but an equal concentration of gases is never reached. Why do you think this is so?

No diffusion here!

There are some places in the human body where substances – rather surprisingly – move from a region of low concentration to a region of high concentration. For instance:

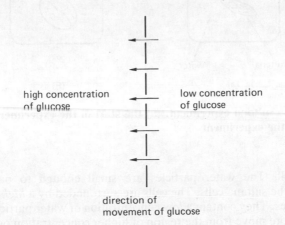

high concentration of glucose

low concentration of glucose

direction of movement of glucose

Movement like this requires a special chemical system. Energy is used. The movement is called *active transport*.

Partial permeability and the big particle

Diffusion can only occur if particles can move freely. In diffusion across a partially permeable membrane, like a cell surface membrane, only small particles can move from one side of the membrane to the other.

Particles of gases like oxygen and carbon dioxide are relatively small and pass easily across partially permeable membranes. Water particles are small too. But cells also contain particles that are too large to pass across the cell surface membrane. So simple diffusion of these particles is not possible.

For example, suppose you put a sultana in a cup of water and leave it overnight. By the next morning it will have swollen. Look at Fig. 2.5.

Exercise

What do you think has happened to the sultana?
Check your answer with that below.

Discussion of results

The sultana appears to have taken up water. We need to take a closer look at the cell surface membrane to understand why this has happened. Look at Fig. 2.6. (This is a simplified diagram of the situation but will help you understand what happens.) Notice that the sultana contains a high concentration of sugar particles. These are big particles. They are too big to pass across the membrane and therefore stay inside the cells. Now look at the water particles inside the cell and

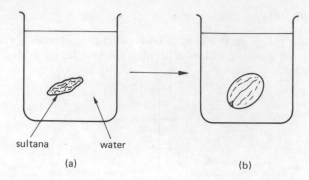

Figure 2.5   The sultana experiment: (a) the start of the experiment, (b) the end of the experiment

outside the cell. The water particles are small enough to pass across the membranes of the sultana cells. The cells are surrounded by a *high* concentration of water particles. They contain a *low* concentration of water particles. The water particles therefore move from the region of higher concentration outside the cells to the region of lower concentration inside the cells.

Theoretically, water would pass into the sultana until the concentration inside was the same as that outside, i.e. until an equal concentration is reached – as in

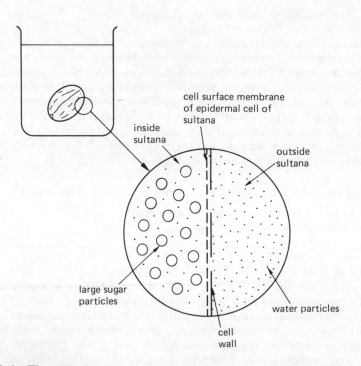

Figure 2.6   The sultana experiment at the membrane level (diagrammatic)

diffusion. This never happens. In fact, the sultana will swell to a certain size and stop. The sultana cells have cellulose cell walls (as well as membranes) which prevent them from bursting.

The sultana experiment is an example of *osmosis*.

┌─ **EXPERIMENT** ─────────────────────────────────
│ *Aim*: To investigate osmosis in eggs
└──

You will need:
2 eggs (same size)
250 cm^3 ethanoic acid or dilute hydrochloric acid
2 × 250 cm^3 beakers
distilled water
200 cm^3 concentrated salt solution
safety spectacles or goggles

Method
1 Put both eggs into a beaker of acid. Leave for 24 hours. This will remove the eggshells (the acid reacts with the carbonate shell). What remains after a day or so is the egg white and yolk surrounded by the cell surface membrane (the egg is a single cell).
2 Place one shell-less egg into a beaker of distilled water.
3 Place the other shell-less egg into a beaker of concentrated salt solution.
4 Leave both eggs for 24 hours.

Questions
1 What has happened to the eggs in this experiment?
2 From your knowledge of osmosis, can you say what has happened to the egg in the distilled water?
3 From your knowledge of osmosis, can you say what has happened to the egg in the concentrated salt solution?

┌─ **EXPERIMENT** ─────────────────────────────────
│ *Aim*: To investigate osmosis in red blood cells
└──

You will need:
a blood sample (do not use human blood)
test tubes
1 cm^3 concentrated salt solution
1 cm^3 salt solution, the same concentration as plasma
2 pasteur (teat) pipettes
microscope slides
coverslips
a microscope

Method
1 Put 5 drops of blood into each test tube. Label the tubes A, B and C.

2

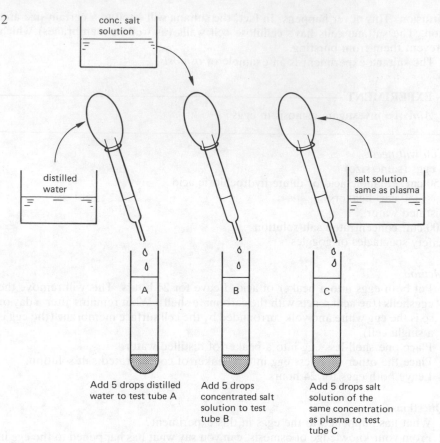

conc. salt solution

distilled water

salt solution same as plasma

A

B

C

Add 5 drops distilled water to test tube A

Add 5 drops concentrated salt solution to test tube B

Add 5 drops salt solution of the same concentration as plasma to test tube C

3 Leave all 3 tubes for 15 minutes. Then put a drop from each test tube on to a microscope slide. Put a coverslip on each slide. Examine the appearance of the red blood cells under the microscope.

4 Record your results in a table like the one below:

Tube	Treatment	Appearance under microscope
A		
B		
C		

Question

Explain the results of the experiment in terms of osmosis.

EXPERIMENT

Aim: To investigate osmosis using Visking tubing

In the laboratory, osmosis is often investigated using Visking tubing or dialysis tubing. It looks and feels like cellophane and is a partially permeable membrane.

You will need:

12 cm Visking tubing
30 cm glass capillary tubing
strong cotton
25 cm³ 10% sucrose solution
distilled water

beaker
retort stand, boss and clamp
pasteur pipette
stopclock

Method

1 Make the tubing wet. Tie a knot in one end, and pipette sucrose solution inside the tubing almost to the top
2 Tie the Visking tubing around the capillary tubing with the cotton.

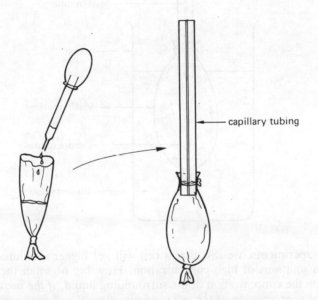

capillary tubing

3 Support the capillary tubing with the clamp on a retort stand.
4 Lower the capillary tubing and Visking tubing into a beaker of distilled water (see the drawing overleaf).
5 Mark the level of the sucrose solution in the capillary tubing.
6 Start the stopclock. At 30 second intervals mark the level of sucrose solution. Continue taking note of the new levels for 15 minutes.
7 Construct a results table for your results.
8 Represent your results graphically.

If you do not know how to construct a results table or how to draw a graph, refer to the Appendix.

Questions

1 From your knowledge of osmosis, give an explanation for your observations.
2 Was the rate of water uptake the same throughout the 15-minute period?
3 If the rate of water uptake did not remain the same, can you suggest a reason?

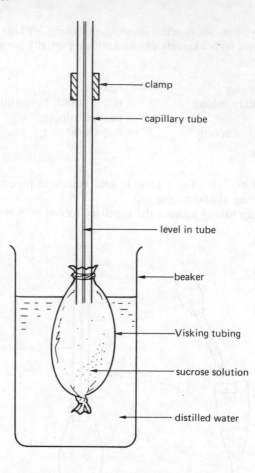

clamp

capillary tube

level in tube

beaker

Visking tubing

sucrose solution

distilled water

In these experiments, we saw that a cell will get bigger in dilute solutions and smaller in solutions of high concentration. How big or small the cell becomes depends on the concentration of the surrounding liquid. *If the surrounding liquid is the same concentration as the cell contents, the cell will not change in size.*

Exercise

Try to write a method for an experiment to determine the concentration inside a potato cell. Write rough notes first – this will enable you to see gaps and errors. Imagine yourself doing the experiment at every stage. Compare your method with the one below. Methods may be different and still fulfil the aim. Some methods may be better and more accurate than others, however.

EXPERIMENT

Aim: To determine the concentration inside a potato cell

You will need:

a potato	kitchen paper
cork borer or sharp knife	balance
ruler	5 test tubes
distilled water	

sucrose solutions: 1 mol dm^{-3}, 2 mol dm^{-3}, 5 mol dm^{-3}, 10 mol dm^{-3}

Method

1 Cut five cylinders of potato, using a cork borer. Cut each cylinder into 3 cm lengths. (If you are using a knife carefully cut five chips, 3 cm long and all the same width.)
2 Rinse the cylinders in distilled water, to remove cell contents from damaged cells.
3 Drain the cylinders on kitchen paper, and blot them dry. Record their masses.
4 Half-fill one of the test tubes with distilled water. Half-fill each of the other test tubes with a different sucrose solution. Label the tubes. Put a potato cylinder into each tube and leave them for one hour.
5 Construct a table for your results.
6 Remove the cylinders. Drain them, and blot them dry. Calculate the change in mass of each piece.
7 Your potato cylinders will all have had slightly different masses. This means that you cannot compare the results for individual pieces directly. So you now need to calculate the *percentage* change in mass for each piece:

$$\% \text{ change in mass} = \frac{\text{change in mass}}{\text{initial mass}} \times 100$$

Add an additional column (% change in mass) to your original results table.

The results will enable you to estimate the concentration in the cells if they are displayed in graphical form. Look at Fig. 2.7. This shows the results from a potato cylinder experiment in a line graph. Find '0' on the vertical axis. The '0' represents no change in mass. Draw a line (marked 'a' on the diagram) from the zero to the plotted line and a second line (line 'b') from the plotted line to the sucrose concentration axis. The point 'c' represents the estimated concentration of the cell.

Summary

- *Diffusion* is the movement of particles from a region of high concentration to a region of low concentration.

- *Osmosis* is the movement of water particles from a region of high concentration to a region of low concentration across a partially permeable membrane.

- Diffusion and osmosis occur in plants and animals.

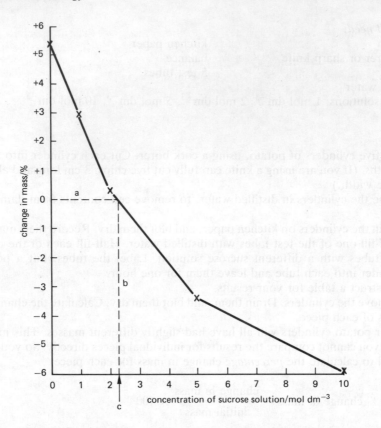

Figure 2.7   Graph to illustrate the change in mass of potato cylinders in sucrose solution: the concentration of sucrose inside the potato cell = 2.3 mol dm^{-3}

Questions on Section 2.1

1 The apparatus is set up as shown in Fig. 2.8.

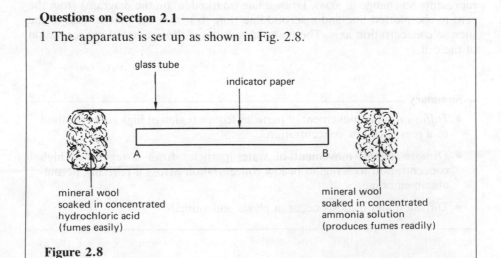

glass tube

indicator paper

A   B

mineral wool
soaked in concentrated
hydrochloric acid
(fumes easily)

mineral wool
soaked in concentrated
ammonia solution
(produces fumes readily)

Figure 2.8

    (a) What do you expect to happen to the indicator at the end labelled A?
    (b) What do you expect to happen to the indicator at the end labelled B?
    (c) Explain why you expect this to happen at A and B. [*Hint*: both acids
       and alkalis change the colour of indicator paper.]
2 Three cylinders of beetroot were prepared. They were all the same
  diameter and the same length. They were then placed in the following:

Cylinder	Solution
A	10% sucrose solution
B	2% sucrose solution
C	distilled water

They were left for 30 minutes.
    (a) Do you think cylinder A would be smaller, bigger or the same size at
       the end of the 30 minutes?
    (b) Do you think cylinder C would be turgid or flaccid?
    (c) Explain your answer to (b).
    (d) Cylinder B hardly changed size at all. Does this tell you anything
       about the concentration of the contents of beetroot cells?

2.2 The cell and chemical reactions

The cell is a living unit, and as such it may be able to grow, reproduce, respire, move, be sensitive and produce excretory products. The chemical reactions taking place in cells enable them to carry out these functions.

Most of the reactions in the body would be very slow, except for substances called *enzymes*. Enzymes are *catalysts* – that is, they speed up chemical reactions without being used up. The particles involved in these reactions can be called *substrate particles*. Enzymes enable substrate particles to react very quickly.

Enzymes are *specific*. Each enzyme has a unique shape and will act as a catalyst for only one type of reaction. You can think of an enzyme as being like a key that will only open one kind of lock. For example, the enzyme pepsin speeds up the breakdown of protein into smaller particles. Pepsin has no effect on lipids or carbohydrate.

Enzymes can work in several ways; for example, they can join together two particles to make a larger one. Look at Fig. 2.9. Here the substrate particles and enzyme particles are moving about randomly. Sooner or later, they will bump into each other or have a collision. Successful collisions will result in the substrate and enzyme joining together. When the enzyme particle has successfully collided with two substrate particles the enzyme comes away, leaving the large particle or product.

Enzymes can also make large particles break down to smaller ones. Look at Fig. 2.10. The large substrate particle and the enzyme are moving about randomly. They will bump into each other, or have a collision. In (b) they have made a successful collision. In (c) the enzyme comes away, leaving the two smaller particles or product.

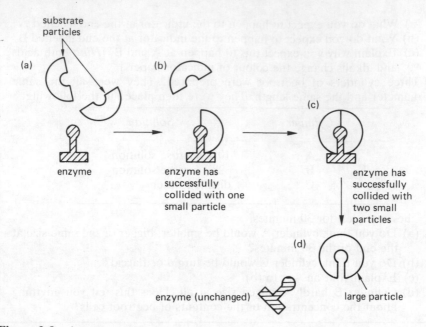

Figure 2.9   An enzyme joining two small particles together to make a large particle

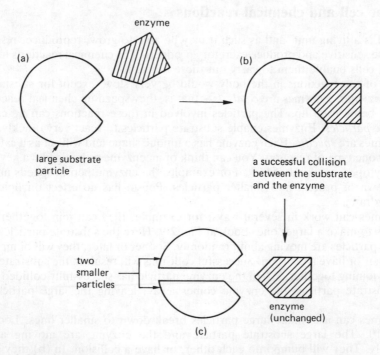

Figure 2.10   An enzyme speeding up the breakdown of a large particle into two smaller-particles

Look at Figs 2.9 and 2.10 once more. Notice that the enzyme and substrate can only fit together in one way. Collisions need to be between the correct points on the enzyme and substrate particle. Think about putting a key into a lock upside down or back to front. That would be an unsuccessful collision.

Enzymes are affected by *temperature*. You already know that enzymes work by colliding with the particles they will join together or break down. If a solution is cold there will be few collisions. As the temperature rises the number of collisions increases. (Read the boxed note 'Heat energy'.) Consequently, there can be more reactions; enzyme action will increase.

Heat energy

Try this experiment to investigate diffusion rates in hot and cold water.

Fill a beaker with hot water, and another beaker with cold water from a refrigerator. Put one drop of ink into each beaker. Leave them to stand absolutely still, but watch them both very carefully.

The ink will diffuse through the water in both beakers. It diffuses at different speeds, however. Diffusion will take place much quicker in the hot water. To understand why this is so, we must consider the particles in the beakers:

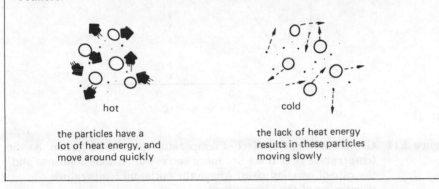

hot

the particles have a lot of heat energy, and move around quickly

cold

the lack of heat energy results in these particles moving slowly

---

There are three main points to remember about enzymes and temperature.

Chemical reactions are speeded up by an increase in temperature and slowed down by a decrease. A rise of 10 °C will double the rate of most of the reactions in the body; that is, they will go twice as fast. A decrease of 10 °C will halve the rate of reaction.

The rate rises until the enzyme reaches its optimum temperature. This is the temperature at which the enzymes work fastest. In human beings, this is between 30 °C and 40 °C. You may be wondering why it doesn't get even faster at temperatures higher than this. This brings us to the third point.

Most enzymes are proteins, and high temperatures disorganise the protein structure. This changes the shape of the enzyme. You saw in Figs. 2.9 and 2.10 that the shape of the enzyme is important. If it is damaged, it cannot work. Think again about an enzyme being like a key in a lock. A damaged key cannot open a lock.

Proteins are said to be *denatured* by high temperatures. The change is irreversible. Egg white, for example, is made up of the protein albumen. As you know, when it is heated it becomes solid, and it can never be changed back into a liquid. An irreversible change has taken place (read the boxed note 'Proteins and temperature above the optimum'.)

Low temperatures slow down chemical reactions, but high temperatures stop enzyme reactions by destroying the enzyme. (Do not describe the enzyme as being 'killed' by heat. An enzyme is not a living thing; it is merely produced by and works in a living cell.) The effect of temperature on the rate of enzyme action can be shown on a graph. Look at Fig. 2.11; notice that as the temperature increases the enzyme is working faster.

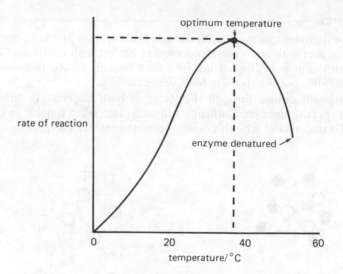

Figure 2.11   Graph showing the effect of temperature on enzyme action. As the temperature rises there are more successful particle collisions and the rate of reaction rises. Above the optimum temperature denaturing of the enzyme begins

Exercise

Look at Fig. 2.11.
At what temperature is the enzyme working best? (That is, what is its optimum temperature?)
Why does the rate fall after it has reached the optimum?

Proteins and temperature above the optimum

Most enzymes are proteins with a very particular shape.

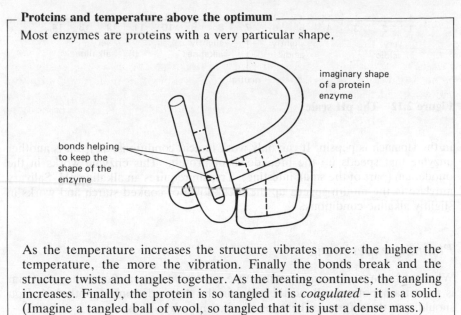

imaginary shape of a protein enzyme

bonds helping to keep the shape of the enzyme

As the temperature increases the structure vibrates more: the higher the temperature, the more the vibration. Finally the bonds break and the structure twists and tangles together. As the heating continues, the tangling increases. Finally, the protein is so tangled it is *coagulated* – it is a solid. (Imagine a tangled ball of wool, so tangled that it is just a dense mass.)

So, the more prolonged the heating the greater the damage, until the structure of the enzyme is so changed that it can no longer function.

Enzymes are affected by acids and alkalis. (If you are not sure what is meant by 'acids and alkalis', do the exercise that follows; otherwise, read straight on.)

Exercise

Consider the following questions. Do not spend more than 10 minutes on this exercise. Write your answers in short notes.

1 (a) What do you think an acid is like?
  (b) Can you think of any naturally occurring acids?
  (c) Can you think of any acids that might be found in the laboratory?
2 (a) What do you think an alkali is like?
  (b) What alkalis do you come across in the home?
  (c) Can you think of any alkalis that might be found in the laboratory?

Compare your notes with those at the end of the chapter.

We say that enzymes are affected by pH. The *pH scale* is used to measure how acidic or how alkaline a solution is. It is shown in Fig. 2.12. The diagram shows that not all acids are equally acidic, nor all alkalis equally alkaline. A liquid with pH 6 would be less acidic than one with a pH 5. A liquid with a pH 8 would be less alkaline than one with a pH 9, and so on.

Enzymes work at an optimum pH. This means that some enzymes can only work in acid conditions, while others can only work in alkaline conditions. For example, the enzyme speeding up the breakdown of protein into smaller particles

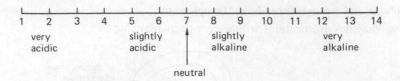

Figure 2.12   The pH scale

in the stomach is pepsin. It can only work in acid conditions. Trypsin is another enzyme that speeds up the breakdown of protein. This enzyme works in the duodenum (part of the small intestine). Trypsin requires an alkaline pH. Salivary amylase in the mouth speeds up the breakdown of cooked starch and works in slightly alkaline conditions.

Planning experiments

When we begin to plan an experiment we must have an aim. Sometimes that aim is to test a hypothesis. Read the notes in the Appendix on 'Hypotheses'. You should note the difference between an aim and a hypothesis.

EXPERIMENT

Hypothesis: The action of rennin is affected by temperature.

Introduction

The experiment uses the enzyme rennin. Rennin is an enzyme found in the stomach of a calf. Cows' milk contains a soluble protein (caseinogen). Rennin will change this soluble protein to an insoluble calcium salt of casein. That is, the liquid milk will clot. Rennin will clot milk in the calf's stomach and in the test tube. You will use rennet (a mixture containing rennin).

[*Teacher's note*: Different batches of rennet give slightly different results. If class results are to be pooled it is advisable to use the same batch of rennet for all experiments.]

You will need:
6 test tubes
thermometer
2 water baths
test tube rack

stopclock
milk (pasteurised)
rennet

Method
1 Set up the water baths at 37 °C and 80 °C.
2 Put 10 cm³ milk into each of three test tubes (labelled A, B and C). Put 1 cm³ rennet into each of the other three test tubes (A_1, B_1 and C_1).

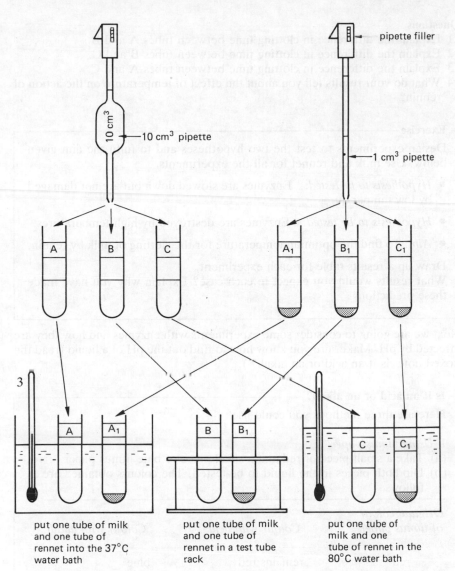

put one tube of milk
and one tube of
rennet into the 37°C
water bath

put one tube of milk
and one tube of
rennet in a test tube
rack

put one tube of
milk and one
tube of rennet in the
80°C water bath

4 Leave for 10 minutes, so that milk and rennet reach the required temperature.
   Record the temperature of the milk and rennet left in the test tube rack.
5 Mix each tube of rennet with the corresponding tube of milk (A_1 with A, B_1
   with B, C_1 with C). Mix thoroughly. Start the stopclock.
6 Examine at intervals of 30 seconds. Record the time taken for the milk to set
   in each tube: record *both* the time at which the milk has just set, although the
   tube cannot be inverted without losing the contents, *and* the time when setting
   is complete and the tube can be inverted without any spillage.
7 Construct your own table that could be used for these results. If you do not
   know to construct a table, see the note in the Appendix, 'Making a table for
   results'.

Questions
1 Explain the difference in clotting time between tubes A and B.
2 Explain the difference in clotting time between tubes B and C.
3 Explain the difference in clotting time between tubes A and C.
4 What do your results tell you about the effect of temperature on the action of rennin?

Exercise

Design experiments to test the two hypotheses and to fulfil the aim given below. Use milk and rennet for all the experiments.

- *Hypothesis to be tested*:   Enzymes are slowed down but are not damaged by low temperatures.

- *Hypothesis to be tested*:   Enzymes are destroyed by high temperatures.

- *Aim*: To find the optimum temperature for the clotting of milk by rennin.

Draw up a results table for each experiment.
What results would you expect in each case? Explain why you have made these predictions.

Next we are going to consider some experiments with enzymes and how they are affected by pH. Make sure you know how to find out the pH of a liquid (read the boxed note 'Is it an acid or an alkali?').

Is it an acid or an alkali?

Here are three methods you could use.

1 *Using litmus paper*
(a) Take a small piece of red and a small piece of blue litmus paper.
(b) Dip both pieces in the liquid to be tested. The colours obtained are as follows:

Original colour of litmus paper	Colour with acid	Colour with alkali
red	remains red	blue
blue	red	remains blue

The acidity or alkalinity of a gas is found by holding wet litmus paper in the gas. Use distilled water to make the paper wet.

2 *Using Universal indicator*
(a) Take a piece of Universal indicator paper.
(b) Dip it into the liquid to be tested.
(c) Match the colour against a colour chart that gives the pH number.

Universal indicator is also available as a liquid. A few drops are added to the liquid to be tested. The colour of the liquid is compared with a colour chart to find the pH number.

3 *Using a pH meter*
A pH meter is an electrical device.

A sensitive probe is placed in the liquid to be tested, and the pH number is read from a display unit. Fig. 2.13 shows a pH meter.

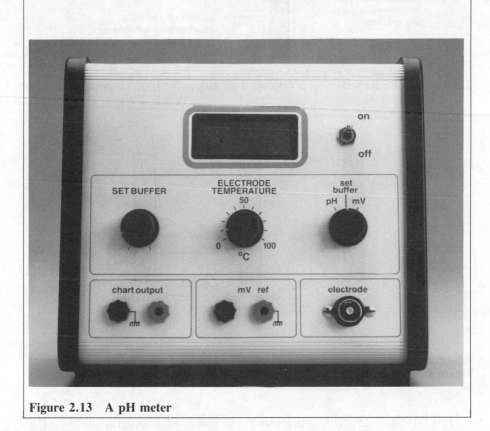

Figure 2.13   A pH meter

EXPERIMENT

Hypothesis to be tested: pH affects the action of the enzyme pepsin

---

Introduction
This experiment investigates the effect of pH on the action of an enzyme. It uses an exposed and developed photographic film.

Look at Fig. 2.14. This represents the structure of the film. Notice that the photographic chemical (shown as dark shading) is stuck to the plastic by protein. If the protein on the film is broken down, it will come away from the plastic. The black photographic chemical will fall away and the plastic will appear clear.

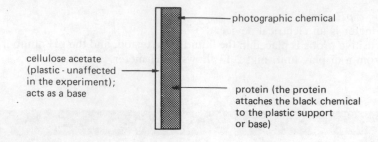

Figure 2.14 Cross-section of a developed photographic film, showing structure

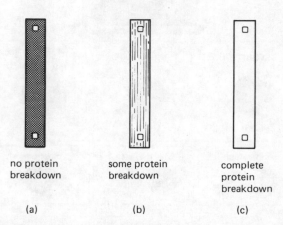

Figure 2.15 Varying degrees of protein digestion on photographic film

Look at Fig. 2.15. This represents three pieces of photographic film. Fig. 2.15(a) shows a piece of film intact. The protein has not been broken down. It looks black due to the photographic chemical.

Look at Fig. 2.15(b). This film looks grey. Some of the black photographic chemical is still attached to the plastic. Not all of the protein has been broken down.

Look at Fig. 2.15(c). This film appears transparent. All the black photographic chemical has come away from the plastic. All the protein has been broken down.

Therefore, the more successful an enzyme is in breaking down protein, the clearer the photographic film. Make sure you fully understand how this works before you continue.

You will need:

3 boiling tubes, labelled A, B and C

water bath (set at 37 °C)

3 stirring rods

3 pasteur pipettes

10 cm³ dilute hydrochloric acid

10 cm³ sodium hydroxide solution

4 pieces photographic film

45 cm^3 pepsin solution
distilled water
3 syringes (or pipettes and
pipette fillers)

Universal indicator solution
stopclock

Method

Safety spectacles and gloves must be worn

1 Put 15 cm^3 pepsin solution and 3 drops of Universal indicator into each boiling tube. Put the tubes in a water bath.
2 Adjust the pH of the tubes: A to pH 7
                                               B to pH ?
                                               C to pH 9
by adding acid or alkali as required.
3 Stir the contents of all three test tubes.
4 Put a piece of photographic film in each tube.
5 Record the appearance of the pieces of film every five minutes for 25 minutes. You will need to construct a table for your results.

Questions
1 Why was the water bath set at 37 °C?
2 Does pH affect the action of pepsin?
3 Does pepsin work best in acidic or alkaline conditions?
4 Pepsin is found in the stomach. What do you think the pH of the stomach is likely to be?

┌─ **Design exercise** ───────────────────────────────────

Hypothesis to be tested: Enzyme concentration affects reaction rate

Design an experiment to test the hypothesis, using rennin (the milk-clotting enzyme). Make a list of the equipment and materials you will need, and draw up a results table.

└──

┌─ **EXPERIMENT** ───────────────────────────────────

Hypothesis to be tested: The size of the substrate particles affects the rate of enzyme action

└──

You will need:
1 cm^3 egg white (from a hard-boiled egg)
1 cm^3 egg white chopped into approximately 8 pieces
1 cm^3 egg white pushed through a fine sieve
3 cm^3 pepsin
dilute hydrochloric acid
5 cm^3 syringe (or pipette and pipette filler)
water bath

Method

Safety spectacles and gloves must be worn

1 Set the water bath at 37 °C.
2 Put 10 cm³ distilled water into each tube, plus the hard-boiled egg white as shown.

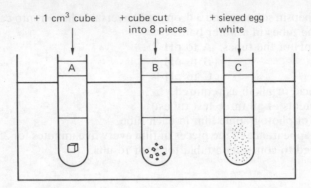

+ 1 cm³ cube   + cube cut   + sieved egg
                    into 8 pieces   white

3 Add 1 cm³ pepsin and 3 drops of hydrochloric acid to each tube. Incubate the tubes in the water bath. Examine them every minute. Record any changes to the pieces of egg white. Construct a table for your results.

Questions
1 Why was the water bath set at 37 °C?
2 Why was hydrochloric acid added to each tube?
3 Does the size of substrate particle affect the rate of pepsin action? Why do you think this is so?

Summary

■ Enzymes speed up chemical reaction.

■ Enzymes are specific.

■ Enzymes are affected by temperature.

■ Enzymes are affected by pH.

■ Most enzymes are proteins.

Questions on Section 2.2

1 You are given the following hypothesis: 'The enzyme amylase, which breaks down starch, is denatured by heat'.
   (a) What does *denatured* mean?
   (b) Give a method that you could use to investigate this hypothesis and a table in which you could record your results.

---

2 Rennet clots milk. It has been suggested that the type of milk may affect the ability of the enzyme to clot it.
  You are provided with pasteurised, longlife (UHT) and raw milk.
  (a) Design an experiment to test this hypothesis.
  (b) Draw up a results table.
3 Salivary amylase is an enzyme which breaks down cooked starch into maltose. Some people think that smoking might reduce the activity of the enzyme. You are provided with samples of saliva from a heavy smoker and a non-smoker, taken during a meal. The mouth was rinsed in water before taking the samples to avoid contamination with food.
  (a) Write a hypothesis connecting smoking with enzyme activity.
  (b) Design an experiment to test this hypothesis.
  (c) Draw up a results table.

Response to exercise

Acids and alkalis (see Section 2.2)
1 You probably found that you knew more about acids than you thought.
  (a) You may have noted that acids

  * are sour
  * burn the skin
  * dissolve away metals
  * are found in oranges and lemons
  * react with substances like chalk
  * kill trees (acid rain).

  (b) Most naturally occurring acids contain carbon. These are called *organic* acids, because they come from or are found in living plants or animals. Here are some examples.

Acid	Where the acid is found
lactic acid	sour milk, yoghurt
ethanoic acid (acetic acid)	vinegar
citric acid	citrus fruits like lemons, oranges and grapefruit
hydrochloric acid (not an organic acid)	your stomach

  (c) The acids commonly used in the laboratory are not organic. They are hydrochloric acid, sulphuric acid and nitric acid.

2 You may have found these questions a little harder to answer.
  (a) You may have said that alkalis

  - feel silky or soapy to the touch
  - can burn the skin.

  (b) Some alkalis that are found in the home:
  bleach
  bicarbonate of soda
  some soaps
  some washing-up liquids
  ammonia in cleaning products, home perm kits, bleaching and tinting kits.

  (c) Alkalis used in the laboratory include sodium hydroxide, potassium hydroxide and calcium hydroxide.

The names of most common alkalis end with the word 'hydroxide'.

3 Making new cells

3.1 Shopping list

Cells in plants and animals do work. In plants, for example, living cells are making food, or transporting food. In animals, they are moving limbs, making secretions, receiving stimuli and so on. Working will result in some cells becoming worn out and these will need replacing. Cells divide to make new cells. New, small cells gradually increase in size, so that they are able to work. To summarise, we can say that cells:

- work
- divide
- increase in size.

To do these things, cells need food.

There are different types of food. Some food provides cells with energy, some enables division to take place. The different types of food are called the *nutrients*.

■ **Memory check**
What are the nutrients the body needs?
What does each nutrient do in the body?
See Section 1.5.

In this section we are going to find out what some of the nutrients are made of. If you do not understand the terms *element*, *atom* and *molecule* read the notes 'Elements, atoms and molecules' in the Appendix.

Carbohydrates

Carbohydrates give us energy. They are made up of the elements *carbon*, *hydrogen* and *oxygen*.

A carbohydrate with small molecules, like *glucose*, can be referred to as a *single sugar*. A glucose molecule has six carbon atoms, six oxygen atoms and twelve carbon atoms. So it is quite large compared to water and carbon dioxide molecules! The atoms in glucose are bonded together in a ring. Look at Fig. 3.1. The structure of glucose is shown in Fig. 3.1(a). Fig. 3.1(b) is a simple representation of the molecule.

Single sugars like glucose are *sweet* and *soluble*, and their molecules are small enough to pass across a partially permable membrane.

If two single sugars bond together, they make a *double sugar*. Look at Fig. 3.2.

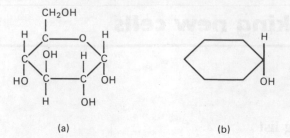

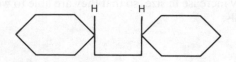

(a)   (b)

Figure 3.1   (a) Structure of a glucose molecule (a single sugar); (b) the structure of glucose in a diagrammatic form

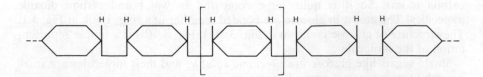

Figure 3.2   Maltose, a double sugar

Notice that the double sugar molecule is simply two single sugar molecules bonded together. Examples of double sugars are:

- *lactose* – a sugar found in milk
- *sucrose* – cane sugar
- *maltose* – a sugar used in brewing beers

Double sugars are also *sweet* and *soluble*, but their molecules are too big to pass through a partially permeable membrane.

The largest carbohydrates are made of many hundreds of single sugar molecules bonded together. They may be bonded in either straight or branched chains. Look at Fig. 3.3. Notice that this carbohydrate is made up of many single sugars.

These 'many sugar' carbohydrates exist in several forms. Here are a few:

- Plants store carbohydrates as *starch*.
- Animals store carbohydrate as *glycogen*. Glycogen is found in the liver and muscles.
- *Cellulose* makes up the cell walls of plants.

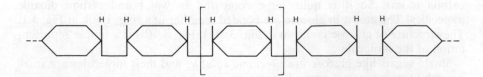

Figure 3.3   Part of a 'many sugar' carbohydrate (the large brackets indicate that this section is repeated many times)

Unlike single and double sugars 'many sugar' carbohydrates are *not sweet*, and they are *not soluble*. They are very large molecules, so they cannot pass through cell surface membranes.

Lipids

Lipids contain the same elements as carbohydrates – *carbon*, *hydrogen* and *oxygen* – but in different proportions from those found in carbohydrates. The atoms are bonded together to form two types of molecule: *fatty acids* and *glycerol*. When these two types of molecule bond together they form a lipid molecule. Look at Fig. 3.4. This represents the structure of a lipid. Notice that the lipid molecule has three fatty acids bonded to one glycerol molecule.

Lipids exist as fats and oils. (*Fats* are solid at room temperatures. *Oils* are liquid at room temperatures.)

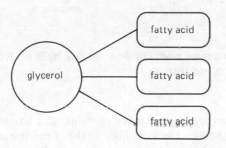

Figure 3.4   Structure of a lipid

Proteins

Proteins are used for growth, and for the repair and maintenance of tissues. They are composed of the elements *carbon*, *hydrogen*, *oxygen*, *nitrogen* and sometimes *sulphur*.

The atoms bond together to make a molecule called an *amino acid*. Not all amino acids are the same. There are 22 different kinds. Look at Fig. 3.5(a). This represents an amino acid. Notice that at one end of the molecule there is an amino group; it contains nitrogen and hydrogen atoms. At the other end there is a carboxyl group; it contains carbon, hydrogen and oxygen atoms.

Every amino acid has an amino group and a carboxyl group. Look at Fig. 3.5(a) again. The part of the molecule separating the amino and the carboxyl groups is different in different amino acids. For example:

- the amino acid glycine has just three atoms between the amino and the carboxyl group
- the amino acid valine has twelve atoms between the amino and the carboxyl group.

Some amino acids are even larger than valine!

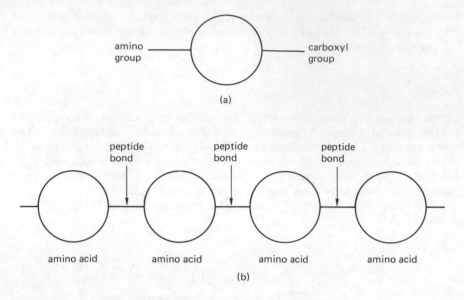

Figure 3.5   (a) An amino acid; (b) proteins are built up from amino acids
bonding together

Amino acids bond together to make proteins. The bonds between amino acids
are called *peptide bonds*. Look at Fig. 3.5(b). Find the amino acids. Find the
peptide bonds. Proteins are large molecules, and may consist of hundreds of
amino acids bonded together.

Amino acids can be bonded together in many different sequences. So many
different types of protein can be made.

EXPERIMENT

Aim: To test foods for single sugars

You will need:
test tubes
test tube rack
scalpel
tile
pestle and mortar
pasteur pipette

beaker
gauze
bunsen burner
range of foods
distilled water
Benedict solution or Fehling solution

Method

1 The diagrams and text below describe how to test both a liquid food and a solid
food for single sugars. The control tube will show what colour the liquid will be
if single sugars are not present. *Safety spectacles and gloves must be worn*.

Testing a liquid food   *Control*   *Testing a solid food (e.g. apple)*

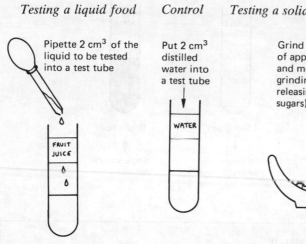

Pipette 2 cm³ of the liquid to be tested into a test tube

FRUIT JUICE

Put 2 cm³ distilled water into a test tube

WATER

Grind a small piece of apple with a pestle and mortar. (The grinding breaks cells, releasing the single sugars)

Add a little distilled water and mix it well with the apple

DISTILLED WATER

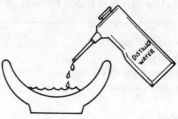

Pour the apple and water into a test tube

APPLE

2 Add Benedict solution or Fehling solution to each test tube.
3 Place the tubes in a beaker of boiling water for a few minutes.
4 Examine the tubes for any change in colour. If the liquid in the tube becomes orange to red, single sugars are present.

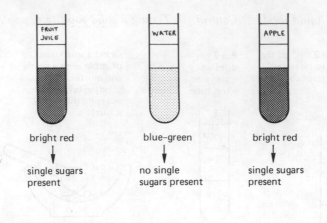

bright red   blue-green   bright red
↓   ↓   ↓
single sugars no single   single sugars
present   sugars present present

EXPERIMENT

Aim: To test foods for starch

You will need:
cavity tile
pasteur pipette
iodine solution

range of foods
scalpel and tile (for cutting solid food)

Method
The diagrams and text below show how to test either a solid food or a liquid food for starch.

1 Put a small piece of the food to be tested into a cavity of a cavity tile. Pipette a few drops of any liquid to be tested into a cavity.

No.	FOOD	COLOUR
1	BREAD	
2	POTATO	
3	RICE	
4	MILK	
5	EGG WHITE	
6	WATER	

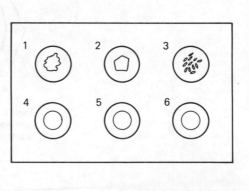

2 Put two drops of iodine solution on to the food in each cavity.

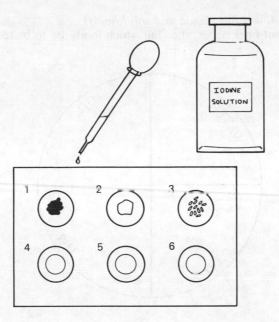

Results

A blue-black colour with iodine shows that starch is present.

A brown colour with iodine shows that starch is not present.

No.	FOOD	COLOUR
1	BREAD	BLUE/BLACK
2	POTATO	BLUE/BLACK
3	RICE	BLUE/BLACK
4	MILK	BROWN
5	EGG WHITE	BROWN
6	WATER	BROWN

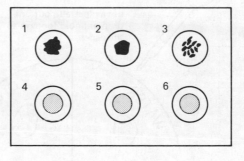

EXPERIMENT

Aim: To test foods for lipids

You will need:

filter paper

pasteur pipettes

test tubes

Sudan III (a red dye)

Method 1 – for oils and fats (liquid and solid lipids)
1 Label a piece of filter paper, showing which foods are to be tested.

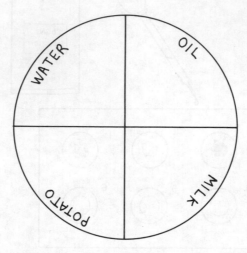

2 Put the food on to the filter paper as shown.

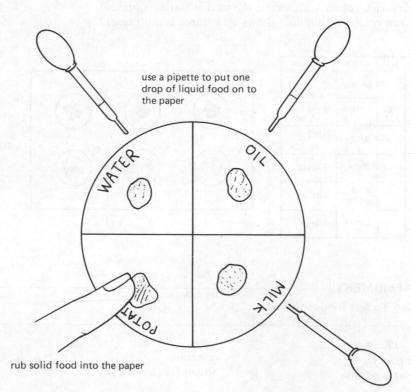

use a pipette to put one
drop of liquid food on to
the paper

rub solid food into the paper

3 Allow to dry.
4 Hold the paper up to the light. A translucent stain shows the presence of a lipid. If a food does not contain a lipid, the food stain is not translucent.

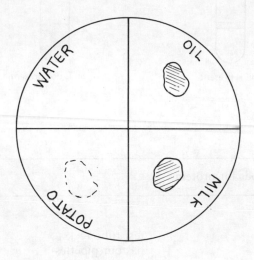

Method 2 – for liquid lipids
1 Add a few drops of oil or the liquid to be tested to a little distilled water.

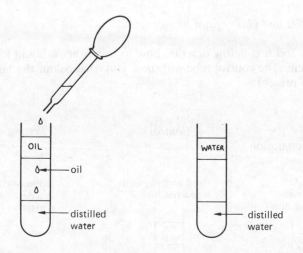

2 Add a little Sudan III (a red dye) to both tubes. Shake the tubes.
3 Allow the tubes to stand. If oil is present, it will settle out on top of the water, stained red. The water remains colourless. If oil is not present, the water remains red; there are no separate layers.

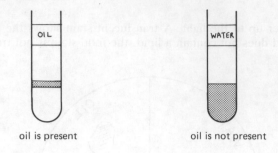

oil is present   oil is not present

EXPERIMENT

Aim: To test foods for protein

You will need:
pestle and mortar
scalpel
tile
test tubes

test tube rack
pasteur pipettes
5% sodium hydroxide solution
1% copper sulphate solution

Method

Safety spectacles and gloves must be worn.

The diagrams and text below describe how to test either a liquid food or a solid food for protein. The control tube will show you what colour the liquid will be if protein is not present.

Testing a liquid   *Control*   *Testing a solid*
  1 Sample preparation

Put 2 cm³   Put 2 cm³ egg white   Grind a little cheese
distilled water   into a test tube   in a pestle and
into a test tube   mortar.

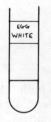

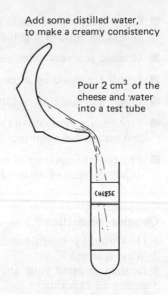

Add some distilled water, to make a creamy consistency

Pour 2 cm³ of the cheese and water into a test tube

CHEESE

2 Add 1 cm³ 5% sodium hydroxide solution to each of the three tubes.
3 Add 1% copper sulphate solution drop by drop to each tube. Shake the tubes after each drop. Look for a colour change.

If protein is present the liquid in the tubes will become purple–violet. If protein is not present, the liquid will be blue–turquoise.

Sometimes biuret solution is used instead of the separate sodium hydroxide and copper sulphate solutions. The test is exactly the same. If biuret is added to a solution containing protein, it will change to a purple–violet colour.

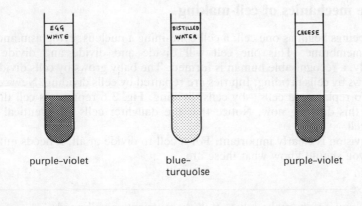

purple–violet   blue–turquoise   purple–violet

── **Summary** ───────────────────────────────
■ *Carbohydrates* are made up of the elements carbon, hydrogen and oxygen.

■ Carbohydrates are made up of many repeating units, e.g. many glucose molecules bonded together.

■ Glucose is a very small carbohydrate molecule.

■ *Lipids* are made up of carbon, hydrogen and oxygen.

■ Lipids are made up of fatty acid molecules bonded to glycerol.

■ *Proteins* are made up of the elements carbon, hydrogen, oxygen, nitrogen and sometimes sulphur.

■ Proteins are made up of many amino acids bonded together. There are 22 different types of amino acid.

Questions on Section 3.1

1 How would you demonstrate that a potato contained starch?
2 What is starch?
3 Bearing in mind your answer to question 2, construct a hypothesis and design an experiment to find out if your answer to question 1 was right. You are provided with cooked mashed potato, an enzyme that can break down starch and any other chemical reagents you think you may need.
2 Consider the following hypothesis: 'Tomato contains more water than lettuce'.
  (a) Design an experiment to test this hypothesis. Make a list of the apparatus you would need, describe what you would do and draw up a results table.
  (b) If you were comparing the water contents of several foods, how could you present your results?

3.2 The mechanics of cell-making

Human beings begin as one cell: a cell containing a nucleus, cytoplasm and a cell surface membrane. This one cell will divide and divide and divide until, eventually, a recognisable human is formed. The baby grows by cells dividing. A child grows by cells dividing. Injuries are repaired by cells dividing. New cells are formed to replace old cells – by cells dividing. Fig. 3.6 represents cell division. Look at this drawing now. Notice that the daughter cells are identical to the mother cell.

Cell division is clearly important. For a cell to divide at all, it needs nutrients; by now you should know what these are.

Exercise

Before you go on, make a list of all the nutrients a cell needs.

Having provided cells with necessary nutrients, we now have to ask – how does a body cell like a skin cell divide? To answer this question, we must concentrate on the nucleus.

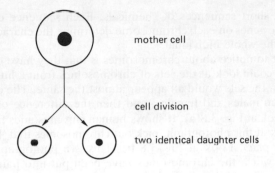

Figure 3.6   Cell division

The nucleus contains a dense material called DNA or deoxyribonucleic acid. This material is in the form of threads within the nuclear envelope. These threads are called *chromosomes*. Look at Fig. 3.7. Notice that the nuclear material is surrounded by sap and then enclosed by a nuclear envelope.

Let's look more closely at chromosomes.

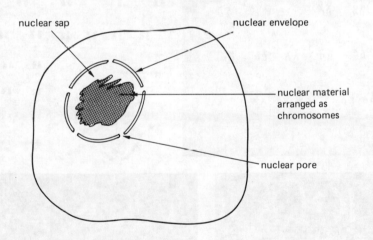

Figure 3.7   Detail of a cell nucleus

Chromosomes

The number of chromosomes in each cell depends on the species of that cell. We say it is *specific*. If you could look at every body cell in a human being (though not sex cells, like sperm and egg cells), you would find 46 chromosomes in each one. This number of chromosomes is found in every skin cell, every muscle cell, every bone cell, every liver cell and so on.

In Section 1.1 you saw that the nucleus contains information. This information is arranged on the chromosomes in the form of a code. The code is composed of a large number of short sequences of chemicals. Each sequence of chemicals is called a *gene*. The genes on each chromosome determine the characteristics of the cell and also of the whole individual.

The next point to notice about chromosomes is that they have definite shapes and sizes. If you could look at the sets of chromosomes from a hundred or more different humans, the sets would all appear almost the same. The only difference would be between males and females, and then the difference only lies in one chromosome. Look at Fig. 3.8(a). It shows human chromosomes from a male. It is possible to look at this photograph, pick out chromosomes that 'look the same' and put them in pairs. Look at Fig. 3.8(b). This is a photograph of the male chromosomes in which the chromosomes have been put into pairs. A chromosome arrangement like this is called a *karyotype*. Fig. 3.9 shows a karyotype being constructed.

Look'at Fig. 3.8(b) again. Find the pairs labelled X and Y. They are different from any other pair. Check this for yourself. The X and Y chromosomes are the

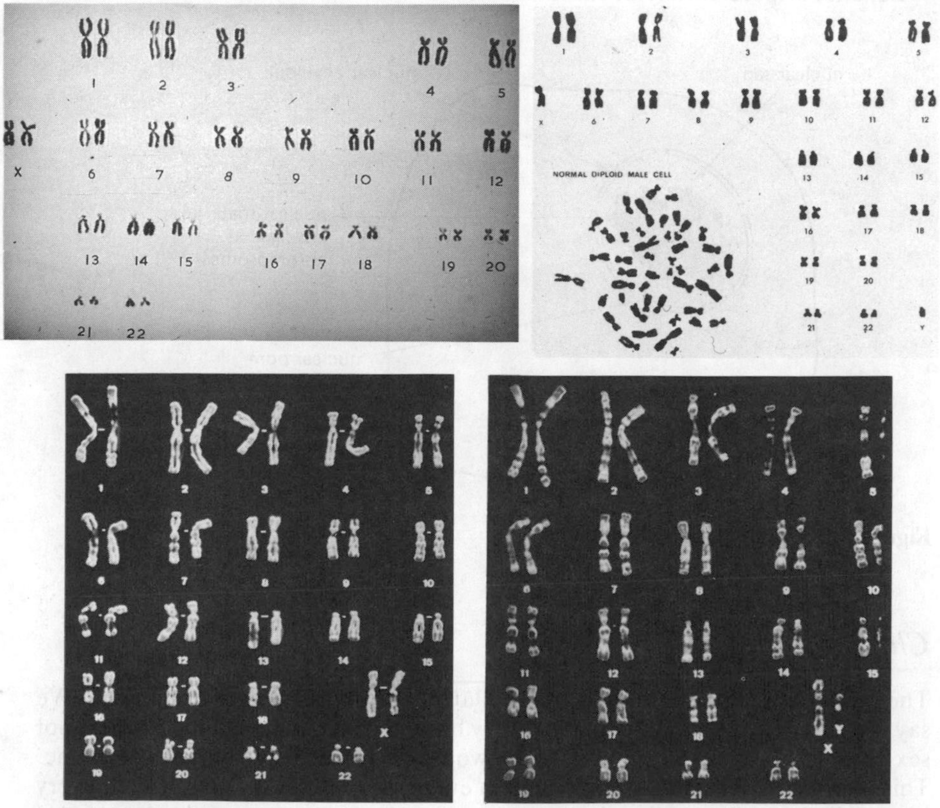

Figure 3.8　Human chromosomes in a cell squash (below, left), and the same chromosomes arranged in homologous pairs

sex chromosomes and are responsible for the sex of the individual. A female has two X chromosomes.

Let us consider one pair of chromosomes. If these chromosomes are exactly the same in length, width and so forth, and make a pair in a karyotype, they are called *homologous* chromosomes. So the two chromosomes labelled '1' in the karyotype (Fig. 3.8(b)) are a homologous pair, the chromosomes labelled '2' are a homologous pair, and so on.

Each chromosome in a homologous pair carries information for the same characteristics. Look carefully at Fig. 3.10(a). (The fruit fly is a convenient organism to use in genetics. The position of its genes have been accurately determined. Consequently, fruit fly chromosome maps have been used here.) Note the bands on the chromosomes. These represent the genes. Note also, that the genes for a particular characteristic are at the *same position* on both chromosomes. Look at the position of the genes on both chromosomes that are responsible for eye colour.

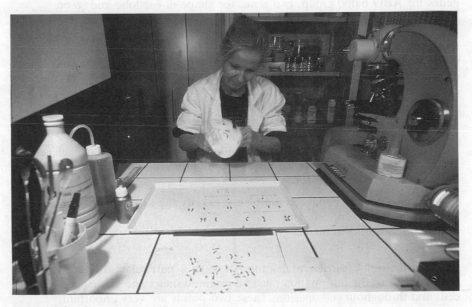

Figure 3.9 Construction of a karyotype

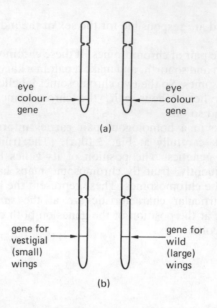

eye
colour
gene

eye
colour
gene

(a)

gene for
vestigial
(small)
wings

gene for
wild
(large)
wings

(b)

Figure 3.10   Two pairs of chromosomes from a fruit fly, indicating the positions of certain genes

Now look at Fig. 3.10(b). Find the genes on both chromosomes that are in the same position, but are different: that is, one gene for small wings and one gene for normal wings. Humans do not have wings but we do have two sets of information for each characteristic we possess. Each human has two genes for eye colour, two genes for ABO blood group, two genes for shape of ear lobe and so on.

Remember that the type of information on each chromosome of a pair may be different; for example one gene may be for brown eyes and the other gene for blue eyes; one gene may be for blood group A and the other for blood group B. It is the interaction of these two sets of information that determines what colour eyes or what blood group we actually have.

Cell division

You can now see that both the number and type of chromosomes in each cell are specific and important. As far as the new cells produced by the division of a cell are concerned, two things follow:

- each daughter cell produced must have exactly the same number of chromosomes as the parent cell, and
- the chromosomes in the daughter cells must be identical to those in the parent cell.

Each chromosome carries vital information in a particular arrangement. The arrangement of the information must be retained intact and unchanged in every cell, and throughout cell division. These two points are very important.

Body cells divide by a sequence of events called *mitosis*. (Sex cells are produced by a different method, discussed in Section 12.3.) Mitosis takes place in the skin, and anywhere else in the body where growth and repair are going on.

Summary

- Cells which are not sex cells divide by a process called *mitosis*.

- All human cells which are not sex cells have 46 chromosomes.

- A cell dividing by mitosis will result in two cells, each with 46 chromosomes.

- The chromosomes in the daughter cells are identical to those in the mother cell.

Questions on Section 3.2

The diagram below shows a simplified cell.

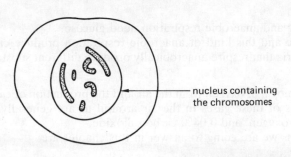

nucleus containing
the chromosomes

1 How many cells would result if this cell divided by mitosis?
2 Draw one of the cells produced after division by mitosis.
3 Where does mitosis take place in the body? Name two places.

4 Energy for life

We learnt in Section 1.5 that all living things respire. Look at the summary of respiration below.

Aerobic respiration

glucose + oxygen ⟶ carbon dioxide + water + energy

Anaerobic respiration (one example)

glucose ⟶ carbon dioxide + alcohol + energy

Note that

- both aerobic and anaerobic respiration need glucose.
- both aerobic and this kind of anaerobic respiration produce carbon dioxide (some bacteria that respire anaerobically produce different substances, such as methane).

Despite the removal of oxygen from the air and the production of carbon dioxide, the proportions of these gases in the air around us are generally more or less constant: 20% oxygen, and 0.04% carbon dioxide.

The questions we are going to answer in this chapter are:

- Where does the glucose come from?
- Where does the oxygen come from?
- Where does the carbon dioxide go to?

4.1 The Sun and the green plant

Imagine yourself carrying out the following experiment.

Hypothesis to be tested: Plants get all their food for growth from the soil

You would need:
10 kg soil
a willow seed
a container

Method
1 Put the soil into a container, taking care not to lose any.
2 Weigh the seed, and plant it in the soil.
3 Water with rain water and keep in a sunny position.
4 Leave for five years.

5 Remove the tree from the container. Collect all the soil from around the roots and from the container.
6 Weigh the tree.
7 Dry the soil and weigh it.

Results

	Initial mass/kg	Final mass/kg
seed	0.001	7.3
soil	10	9.9

Conclusions
The plant has increased its mass by 7.299 kg. The soil has lost 0.1 kg in mass. It looks as though most of the increase in mass in the plant must be due to the water.

This experiment was carried out in the early sixteenth century by a Flemish physician, Van Helmont. The results are not his but his conclusion was that the material of the tree came from the water.

Exercise

Do you agree with Van Helmont's conclusion? If you don't, how could you show him that the increase in mass was not just due to the water in the tissues?
Spend no more than five minutes thinking about this. Compare your suggestion with the one below.

If you dried the tree to remove all the water, you would be left with material that exceeded the mass of the seed. Where has this extra mass come from? In 1727 an English botanist, Hales, suggested that the extra mass came from the air. Understanding how green plants make their food, and therefore grow, has taken over 200 years of countless hypotheses and experiments.

We now know that if a green plant is given light, carbon dioxide and water it is able to make glucose and oxygen. This can be represented in a sum or an equation:

carbon dioxide + water + light $\longrightarrow$ glucose + oxygen

Look at the equation carefully. It shows how the plant takes in small molecules from its surroundings and uses them to make larger molecules. That is, it takes in carbon dioxide and water and makes carbohydrates like glucose.

■ **Memory check**
What are the elements in carbohydrates?
What is the difference between glucose and starch?
See Section 3.1.

Let's look at how the plant makes these carbohydrate molecules. The process is called *photosynthesis*.

The process of photosynthesis

1 Light is vital for photosynthesis. Plants are designed to get as much light as possible.

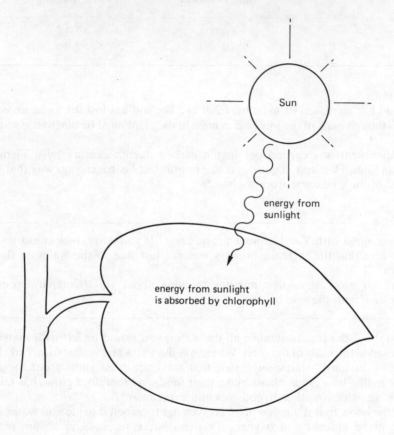

Sun

energy from
sunlight

energy from sunlight
is absorbed by chlorophyll

■ **Memory check**
What is chlorophyll?
Where would you find chlorophyll in the cell?
See Section 1.1.

2 Water enters the plant by the roots and travels to the leaf in water-carrying vessels.

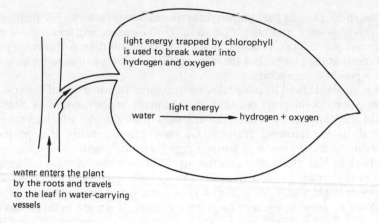

This diagram shows what happens to the breakdown products of water:

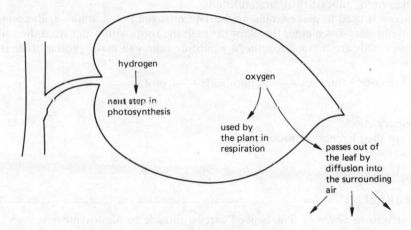

3 Carbon dioxide diffuses into the leaf and reacts with the hydrogen.

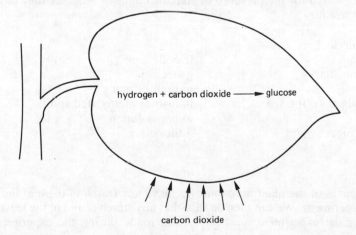

You can think of cells that photosynthesise as small factories. A factory will store its products until transport is possible. Good storage will mean more room for continued production. Glucose molecules could build up to form very high concentrations in the plant. They therefore quickly bond together to make starch. Starch is a good storage product.

Starch is insoluble, and its molecules are too large to move out of the cell. But the growing parts of the plant need the glucose made in photosynthesis. Starch in the cells of the leaf must be broken down into soluble glucose, which can then be transported to the growing regions. Glucose can be easily transported in food-carrying tubes to growth or storage regions in the plant.

Think back to Van Helmont's experiment. Photosynthesis is really only part of the story of the growing willow tree. To grow, plants need protein. Let's look at what happens to glucose. Look at Fig. 4.1. Glucose is transported up the growing shoots, down to growing roots or to storage organs. It is used in three ways.

- Glucose is used as a source of energy.
- Glucose is stored for future use in organs such as fruits, seeds, stems, leaf stalks, roots, tubers, rhizomes and bulbs.
- Glucose is used to make amino acids. The nitrogen in the amino acids comes from nitrates, which enter the plant through the roots with water from the soil. Amino acids are bonded together by peptide bonds to make protein; that is

$$\text{glucose} + \text{nitrates} \longrightarrow \text{amino acids} \longrightarrow \text{protein}$$

■ **Memory check**
What are the elements in protein?
See Section 3.1.

┌ EXPERIMENT ─────────────────────────────────

Hypothesis to be tested: Plants need carbon dioxide to photosynthesise

Leaves are tested for the presence of starch to indicate whether they have been photosynthesising.

You will need:

potted plant	tripod
2×100 cm^3 flasks	gauze
soda lime	bunsen burner
non-absorbent cotton wool	industrial methylated spirits
beaker	iodine solution
3 boiling tubes	3 tiles
forceps	

Method
1 The leaves of the plant used must not have any starch in them at the start of the experiment. We can then be sure that any starch found in the leaves at the end of the experiment must have been made during the experiment. The

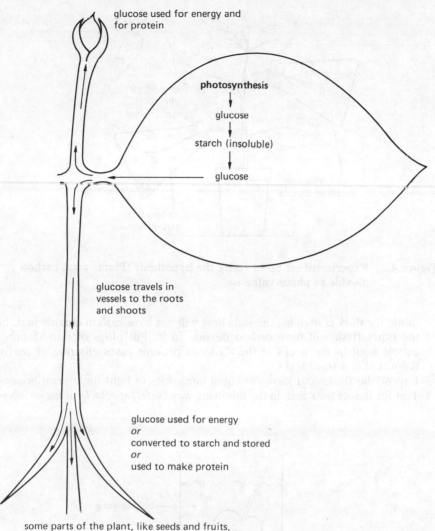

glucose used for energy and
for protein

photosynthesis

↓

glucose

↓

starch (insoluble)

↓

glucose

glucose travels in
vessels to the roots
and shoots

glucose used for energy
or
converted to starch and stored
or
used to make protein

some parts of the plant, like seeds and fruits,
may convert glucose into lipids

Figure 4.1   What happens to glucose in the plant

starch is removed by keeping the plant in a dark cupboard for 48 hours. All the
starch in the leaves will be converted into glucose and moved to growing areas
or storage organs. This process is called *destarching*.

2 Remove the plant from the cupboard and immediately set up the experiment
shown in Fig. 4.2. Enclose one leaf of the plant, labelled A, in a conical flask
containing a few grams of soda lime. This will absorb carbon dioxide from the
air around the shoot. Enclose a second leaf (label it B) in a similar flask. This
is the control flask – it copies the experimental flask exactly except for one

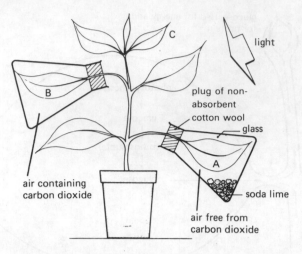

Figure 4.2   Experiment set up to verify the hypothesis 'Plants need carbon dioxide to photosynthesise'

thing: the flask containing the soda lime will not have carbon dioxide in it, but the other flask will have carbon dioxide in it. Put plugs of non-absorbent cotton wool in the necks of the flasks to prevent gases entering or leaving them. Label a third leaf C.

3 Expose the flasks and leaves to equal intensities of light for several hours.

4 Test the leaves for starch in the following way (*safety spectacles must be worn*):

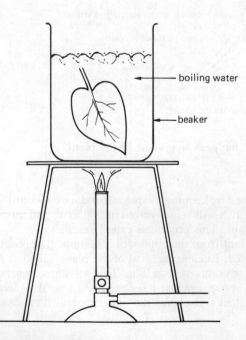

(a) Pick the labelled leaves, and kill each one by boiling for 10 seconds. Boiling stops all chemical activity in the leaf and will make the leaf more permeable to iodine later.
(b) *Turn off the bunsen burner before you go on.* Put each leaf into a boiling tube and cover with methylated spirits. Put the boiling tubes into the beaker of hot water.

Care! Methylated spirits is highly flammable.

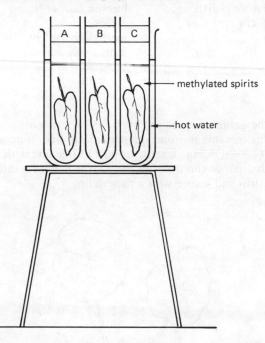

The chlorophyll will dissolve out of the leaf and into the methylated spirits. Check the leaves after five minutes to see if their colour has gone.
(c) When the leaves are white, tip the methylated spirits into a beaker. Fill the tubes with water. The leaves should float to the top of the tube. Lift a leaf out with the forceps. It will be very crisp. This is an effect of the methylated spirits. Rinse each leaf in hot water. This will soften the leaves.
(d) Spread the leaves on tiles and flood them with iodine. A blue-black colour with iodine indicates the presence of starch. A brown colour with iodine indicates the absence of starch.
5 Construct a results chart for your results.

Questions
1 Why were the results for leaves A and B different? Does this result verify the hypothesis?
2 Why were the results for leaves B and C different?
3 Why do you think vegetable growers enrich the air inside greenhouses with carbon dioxide?

> ⚠️ **EXPERIMENT** ──────────────────
>
> *Hypothesis to be tested*: Light is necessary for starch production in leaves

You will need:

geranium plant	forceps
strip of aluminium foil,	tripod
12 cm × 1.5 cm	gauze
strip of transparent polythene,	bunsen burner
12 cm × 1.5 cm	2 tiles
2 paper clips	methylated spirits
beaker	iodine solution
2 boiling tubes	

Method
1 Destarch the geranium as in the previous experiment.
2 Remove the geranium from the cupboard and immediately set up the experiment shown in Fig. 4.3. Cover part of one leaf with the aluminium foil. Secure with a paper clip if necessary. Cover part of a different leaf with the polythene strip and secure with a paper clip.

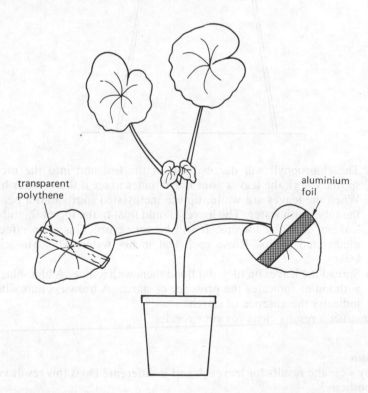

transparent
polythene

aluminium
foil

Figure 4.3   Experiment set up to verify the hypothesis 'Light is necessary for starch production'

3 Test both leaves for starch, as in part 4 of the previous experiment. Remove the foil and polythene coverings before boiling the leaves.
4 Construct a table for your results.

Questions
 1 Consider the leaf that was covered with aluminium foil. Does the appearance of this leaf indicate that light is necessary for starch production? Explain your answer.
 2 Why was a second leaf covered with polythene?
 3 If part of a leaf is covered with aluminium foil, it is deprived of light. Could it be deprived of anything else?

EXPERIMENT

Hypothesis to be tested: Starch is produced in areas of a leaf that contain chorophyll

You will need:

a variegated plant	boiling tube
a beaker	forceps
tripod	tile
gauze	methylated spirits
bunsen burner	iodine solution

Method
 1 Keep the variegated plant in sunlight for several hours.
 2 Detach one leaf. Draw it carefully. This is a *chlorophyll map*.
 3 Test the leaf for starch as in the previous experiments.

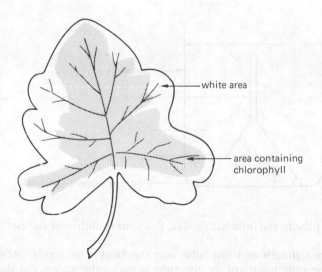

Figure 4.4   Chlorophyll map of a leaf

4 Draw the leaf after it has been flooded with iodine. This is a *starch map*.
5 Compare the two drawings.

Questions
1 Which areas of the leaf contained starch?
2 Is starch produced in areas that contain chlorophyll?
3 Why do you think variegated plants are never found in natural forests?
4 If the areas without chlorophyll do not photosynthesise, how do they get food to grow?
5 A student thinks that maybe the areas without chlorophyll produce glucose instead of starch. Write a hypothesis that she could test by doing an experiment. What apparatus would she need, and how would she do the experiment?

┌─ **EXPERIMENT** ──
│ *Hypothesis to be tested*: Oxygen is given off from a plant while it is in light
└───

You will need:
2 sprigs of *Elodea* (pond weed)   pond water
1 dm³ beaker   small piece Plasticine
glass funnel   wooden splint
test tube

Method
1 Place the pond weed under a funnel in a large beaker of pond water (if you have no pond water, use water with a little sodium carbonate added).

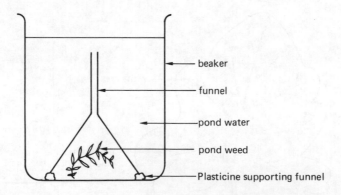

2 Fill a test tube to the brim with water. Put your thumb over the end. Invert the tube.
3 Lower your thumb and test tube into the beaker of water. Remove your thumb when the mouth of the test tube is under the water. Put the test tube over the funnel as shown.

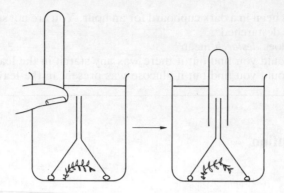

4 Shine a lamp on the apparatus and look for bubbles rising from the plant.
5 The bubbles will rise up the funnel into the test tube of water and displace the water from the tube.
6 When the tube is about half-full of gas, lift the tube carefully from the funnel. Put your thumb over the mouth of the test tube while it is still in the water. Remove the tube and light a wooden splint. Blow it out so that it is glowing red at the end. Plunge this into the gas in the test tube (not in the water). If oxygen has been collected, the glowing splint will burst into flames.

Questions
1 Is oxygen given off from a plant when it is in the light?
2 What process is producing this oxygen?
3 Do you think the amount of oxygen given off will be affected by the strength of the light given to the plant? Write a hypothesis for an experiment that would investigate this. What apparatus would you need? Construct a results table. Could you present your results in a way that would illustrate the relationship between the strength of the light and oxygen production?

[*Hint*: Remember that 40, 60 and 100 watt bulbs give light of different strengths.]

Summary

■ The term *photosynthesis* describes how green plants use radiant energy from the sun to make food from simple molecules.

■ To photosynthesise, a green plant needs *carbon dioxide, water, chlorophyll* and *light*.

■ The product of photosynthesis is *glucose*, which is then converted into *starch*.

Questions on Section 4.1

1 (a) Why do market gardeners pump carbon dioxide into their greenhouses?
  (b) Why do they illuminate them at night time?
2 (a) At what time of the day are plants producing the most oxygen?
  (b) Why is this so?

3 A plant has been in a dark cupboard for an hour. You are not sure whether it has been destarched.
  (a) What does *destarch* mean?
  (b) How could you find out if there was any starch in the leaves?
  (c) How could you find out if glucose was present in the leaves?

4.2 **Deforestation**

Forests

We know from Section 4.1 that plants photosynthesise. This process provides us with oxygen and food. A forest consists of many plants and can be thought of as a huge photosynthesising system. But in many parts of the world vast areas of forest are being destroyed.

What is deforestation?

Trees and other plants making up a forest are cut down and/or burnt. The ground is cleared. In 1950, 30% of the Earth's surface was covered by tropical forest; by 1975 this proportion had fallen to 12%. Forests are cleared both by farmers and commercial loggers.

Farmers, particularly in third world countries, practise 'slash and burn'. The forest is cut down by machete and burnt. Look at Fig. 4.5. This shows 'slash and burn' in Cameroon, West Africa.

Commercial loggers use machines. A 'tree crusher' will pull down and pulp several large forest trees in an hour. The chipper reduces trees to chips smaller than a one-penny piece. It can consume one tree every minute. It can clear 12 acres of forest in a day. Look at Fig. 4.6. This shows loggers and machinery clearing forest.

Why does all this happen? Read the boxed note 'Why the forests are cleared'.

Why the forests are cleared

Farmers clear forest areas so that:

- land is available for farming; good fertile soil is often used for cash crops like coffee, bananas, sugar or rice, but farmers also have to clear forests to grow their own food crops, even though forest soil is not very fertile and can only be used for a short time
- they can obtain firewood as a fuel for cooking.

Commercial loggers will clear forest areas so that:

- roads can be made
- space is available for cattle ranches – two-thirds of the forests in Central America have been cleared for this purpose
- hardwoods like mahogany can be obtained
- wood can be sold and the money used to pay national debts.

Figure 4.5 Slash and burn in Cameroon, West Africa

Are forests important?

You can answer this question for yourself when you have finished reading this section.

Look at Fig. 4.7. Notice how many of our basic living requirements come from trees. Now look at Fig. 4.8. The forests influence the environment in which we live. Try the exercise below which will help you to understand the importance of the forests, and whether deforestation is a problem.

Exercise

Copy Fig. 4.8, but draw the stumps of the trees instead of the forest. Put notes on your drawing describing the effects of removing the forest from the hillside. The questions below will help you.

1 If trees are cut down, what will happen to levels of oxygen and carbon dioxide in the atmosphere?

2 If trees are burnt, what will happen to levels of oxygen and carbon dioxide in the atmosphere?
3 The sun is shining. Will the ground on the hillside be hotter or colder when the trees are removed?
4 The sun is shining, it is windy and dry. What could happen to the soil on the hillside?
5 It rains very hard. What will happen to the soil on the hillside?
   What will happen to the crops at the foot of the hillside?
   What will happen to the homes at the foot of the hillside?
   What will happen to the lake?
6 What could happen to the climate of this area if the trees are removed?
7 What will happen to the plants and animals living in the forests?

Figure 4.6   Loggers and machinery clearing forest in Indonesia

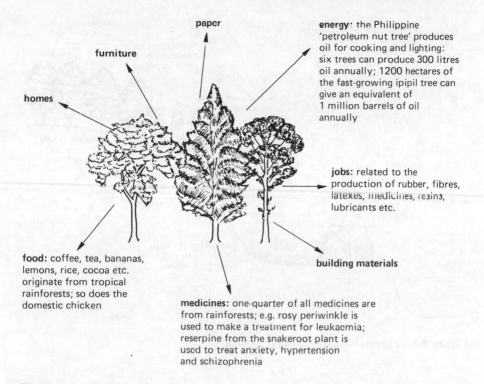

paper

furniture

homes

energy: the Philippine 'petroleum nut tree' produces oil for cooking and lighting: six trees can produce 300 litres oil annually; 1200 hectares of the fast-growing ipipil tree can give an equivalent of 1 million barrels of oil annually

jobs: related to the production of rubber, fibres, latexes, medicines, resins, lubricants etc.

food: coffee, tea, bananas, lemons, rice, cocoa etc. originate from tropical rainforests; so does the domestic chicken

building materials

medicines: one-quarter of all medicines are from rainforests; e.g. rosy periwinkle is used to make a treatment for leukaemia; reserpine from the snakeroot plant is used to treat anxiety, hypertension and schizophrenia

Figure 4.7 Forests – providers

┌─ **Summary** ─────────────────────────────────

■ *Deforestation* is the clearing of trees from forests.

■ Forests provide homes for people, animals and plants.

■ Forests provide us with oxygen and remove carbon dioxide from the atmosphere.

■ Forests provide us with food and many other useful materials.

■ Removing forests has disastrous consequences on the environment and climate.

┌─ **Questions on Section 4.2** ─────────────────────────────────

1 The hills around the Panama Canal are being cleared of their trees by local farmers. The lake which supplies the canal with water is silting up. Do you think there could be a link between these two observations? Explain your answer.

2 The organisation Friends of the Earth believe that by not eating beef-burgers and not buying hardwoods like mahogany (for furniture) we can help to save the rainforests. Why do they think so?

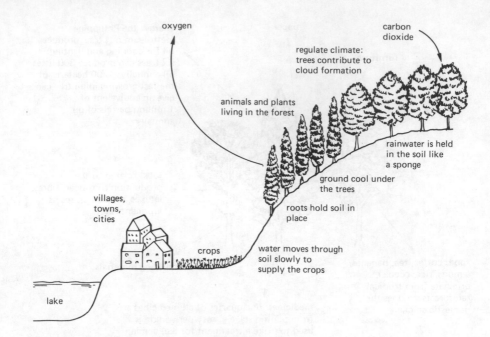

Figure 4.8   Forests – environmental influence

4.3 The greenhouse nobody wants

The last ice age ended 10 000 years ago. Since then, the Earth has warmed by about 5 °C. Scientists predict that the temperature could rise between 2 °C and 8 °C in the next 100 years. The world is getting hotter, and it is getting hotter very quickly.

The warming is due to high levels of 'greenhouse gases' in the layer of air around the Earth (the stratosphere). The most important of these gases are:

carbon dioxide   nitrogen oxides
chlorofluorocarbons ('CFCs')   ozone
methane

Look at Fig. 4.9(a). Notice that the energy from the Sun enters the Earth's atmosphere. Most of it is lost by reflection.

Now look at Fig. 4.9(b). This is what is happening to our world at the moment. The energy from the Sun enters the Earth's atmosphere. The Earth reflects it, but it cannot pass back into space through the layer of greenhouse gases that has built up.

So where have these greenhouse gases come from? Study Table 4.1 carefully. Notice that ozone has quite a complicated part to play – read the boxed note 'Ozone'. Now answer the questions in the exercise.

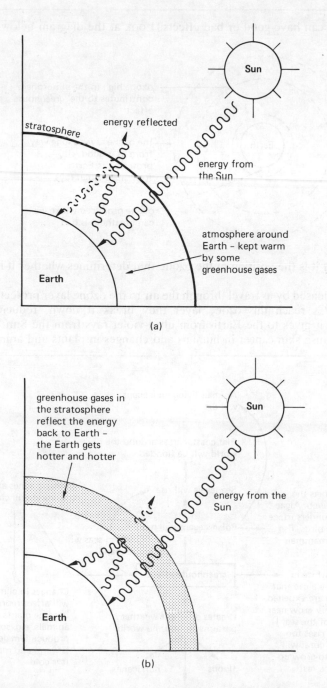

Figure 4.9   (a) The Sun's energy is reflected from the Earth; (b) the greenhouse effect

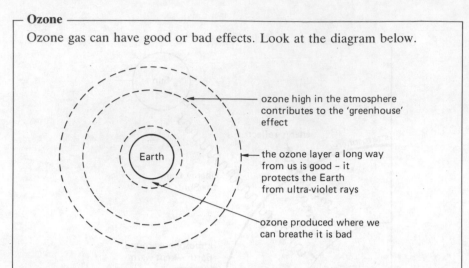

Ozone

Ozone gas can have good or bad effects. Look at the diagram below.

Notice that it is the position of the ozone that determines whether it is good or bad.

CFCs released by us travel through the air to the ozone layer protecting us. When CFCs reach this ozone layer they break it down, reducing the protection it gives to the Earth from ultra-violet rays from the Sun. These rays can cause skin cancer in humans and changes in plants and animals.

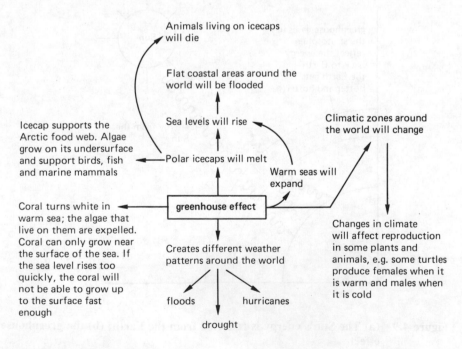

Figure 4.10   The greenhouse effect

Table 4.1 Sources of greenhouse gases

Gas	Source	Other effects
carbon dioxide	combustion of trees and fossil fuels (this includes oil- and coal-fired power stations); respiration (carbon dioxide levels in the atmosphere have steadily risen for many years)	
chlorofluorocarbons (CFCs)	some aerosols; refrigerants; some foamed plastics	break down the ozone layer that surrounds and protects the Earth
methane	decomposition of animal and vegetable remains	
nitrogen oxides	combustion of fossil fuels; manufacture and use of artificial fertilisers	contribute to acid rain
ozone	reaction between carbon monoxide, methane and other hydrocarbons and nitrogen oxides in sunlight (e.g. car exhaust fumes on sunny days) (photochemical smog)	irritates lungs of people with breathing disorders like asthma; plant growth slows down

---

Exercise

What can be done to reduce the levels of carbon dioxide produced?
What can be done to increase the removal of carbon dioxide from the atmosphere?
How could the production of other greenhouse gases be reduced?

Scientists do not understand exactly what will happen as the greenhouse effect increases. They can only guess, and judge by what already is happening. Fig. 4.10 represents some of the effects of global warming. Study these carefully.

Summary

■ The *greenhouse gases* are carbon dioxide, chlorofluorocarbons, methane, nitrogen oxides and ozone.

■ The gases stop the energy from the Sun from being reflected back into space.

■ Global warming will cause sea levels to rise, weather patterns and climate to change, and possibly many other serious effects.

PROJECT WORK

Keep a file of cuttings from newspapers and magazines that mention the greenhouse effect. Keep sections of your file on

● incidents thought to be due to the greenhouse effect

● predictions – what could happen as a result of the greenhouse effect

● methods of reducing the greenhouse effect

● surviving the greenhouse effect; for example, in 1988–89 'conservation corridors' were being discussed in America, to allow wildlife to move with the changing climate.

4.4 **Fossil fuels and dying forests**

The combustion of fossil fuels produces oxides of both sulphur and nitrogen.

Exercise

Use the information in the table below to draw a bar chart representing the percentage contributions of various sources of sulphur dioxide and nitrogen dioxide emissions.

Source	Sulphur dioxide	Nitrogen dioxide
domestic heating	5	4
commerce and industry	18	21
vehicles	11	29
power stations	66	46

The dry particles produced in combustion can fall to the ground near to where they are produced. This is *dry deposition*. The effects of dry deposition are shown in Fig. 4.11.

■ **Memory check**
What is an acid?
What is pH?
What is the pH range for acids?
See Section 2.2.

The particles of acid dust can dissolve in water droplets in the atmosphere and come down to the ground in rain, snow, mist or fog. This is *wet deposition*. Look

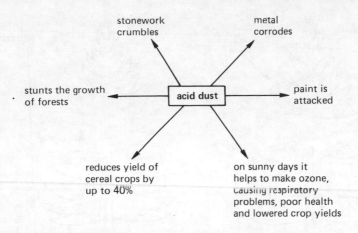

Figure 4.11   The effects of dry deposition

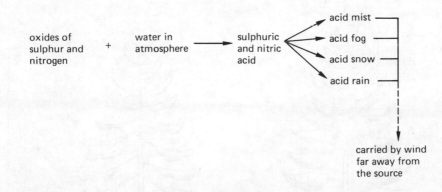

Figure 4.12   The formation of wet deposition

at Fig. 4.12. This summarises the formation of wet deposition. Clearly, wet deposition lands on soil, plants, lakes and rivers, as well as buildings. Look at Fig. 4.13 (pages 88–89). This shows how wet deposition affects trees and soil. Trees starve, due to the lack of nutrients, and then become so weak that they are easily attacked by fungi, insects and other pests. The damage to forests is increasing.

Look at Fig. 4.14 (pages 90–91). Notice that the metals released in the soil due to wet deposition are leached through the soil and may enter rivers and lakes. Take note of the ways in which these metals affect life in and around the rivers and lakes. Notice that water pumped out of a lake for drinking may contain toxic metals. Some scientists think that aluminium in drinking water may be related to senile dementia (also known as Alzheimer's disease). This disease has become more common in parts of Norway where acid rain has destroyed forests and affected lakes.

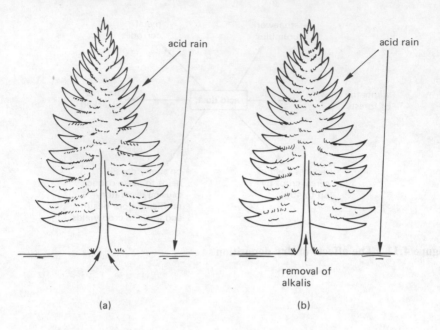

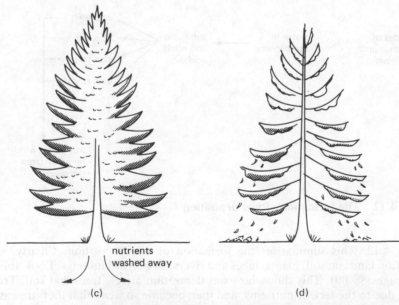

Figure 4.13   The effect of wet deposition on trees and soil: (a) alkaline salts are taken up by the trees to neutralise the acid; (b) the acid dissolves nutrients out of the soil, releasing poisonous metals; (c) lacking nutrients, the leaves turn yellow; (d) the yellow leaves die and fall; with few leaves to photosynthesize less food is made (Figure 4.13 continues on page 89)
continues on page 89)

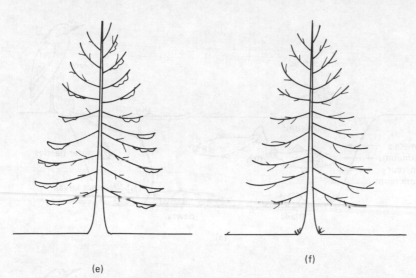

(e)

(f)

Figure 4.13 *(continued)* **(e) with too little food, the trees become weak; new branches cannot develop properly, and diseases can attack easily; (f) death of the trees**

Exercise

The Black Forest has been severely damaged by acid rain. In many tourist areas around and in the forest, damaged sick trees have been cut down and new trees planted. Do you think this is a solution to acid rain?

Solutions

The problems of acid rain could be solved. Look at Fig. 4.15.

Exercise

1 Two of the ways of solving the acid rain problem involve the use of limestone. Find the methods that use limestone in Fig. 4.15.
  (a) Flue-gas desulphurisation would use vast amounts of limestone if it were installed in coal-burning power stations. Find out where limestone comes from.
  (b) Could supplying the limestone create environmental problems?
  (c) The calcium sulphite formed at first becomes converted to calcium sulphate. What do you think will happen to all the calcium sulphate produced? Find out if calcium sulphate can be used for anything.
  (d) Do you think this method of solving acid rain is a good one?
2 Make a list of all the ways you and your family can help to reduce acid rain.

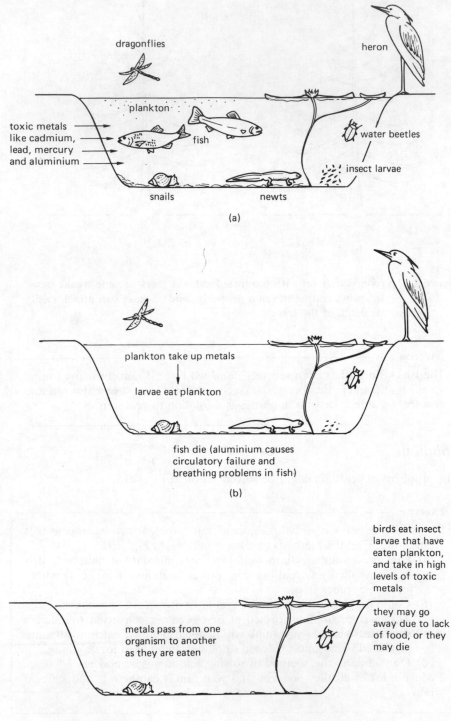

dragonflies

heron

plankton

toxic metals
like cadmium,
lead, mercury
and aluminium

fish

water beetles

insect larvae

snails

newts

(a)

plankton take up metals

larvae eat plankton

fish die (aluminium causes
circulatory failure and
breathing problems in fish)

(b)

birds eat insect
larvae that have
eaten plankton,
and take in high
levels of toxic
metals

they may go
away due to lack
of food, or they
may die

metals pass from one
organism to another
as they are eaten

(c)

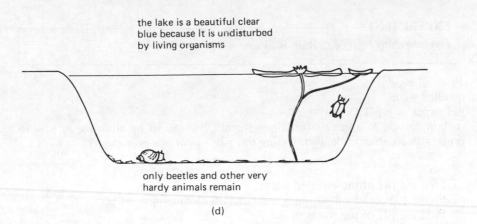

the lake is a beautiful clear
blue because It is undisturbed
by living organisms

only beetles and other very
hardy animals remain

(d)

Figure 4.14   The effects on lakes and rivers of metal salts released as a result of wet deposition

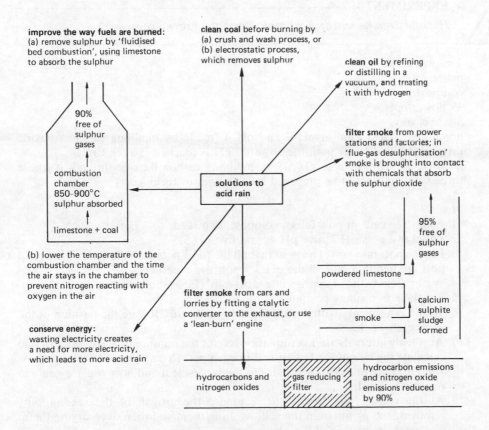

improve the way fuels are burned:
(a) remove sulphur by 'fluidised
bed combustion', using limestone
to absorb the sulphur

clean coal before burning by
(a) crush and wash process, or
(b) electrostatic process,
which removes sulphur

clean oil by refining
or distilling in a
vacuum, and treating
it with hydrogen

90%
free of
sulphur
gases

combustion
chamber
850–900°C
sulphur absorbed

limestone + coal

filter smoke from power
stations and factories; in
'flue-gas desulphurisation'
smoke is brought into contact
with chemicals that absorb
the sulphur dioxide

**solutions to
acid rain**

95%
free of
sulphur
gases

powdered limestone

calcium
sulphite
sludge
formed

smoke

(b) lower the temperature of the
combustion chamber and the time
the air stays in the chamber to
prevent nitrogen reacting with
oxygen in the air

filter smoke from cars and
lorries by fitting a ctalytic
converter to the exhaust, or use
a 'lean-burn' engine

conserve energy:
wasting electricity creates
a need for more electricity,
which leads to more acid rain

hydrocarbons and
nitrogen oxides

gas reducing
filter

hydrocarbon emissions
and nitrogen oxide
emissions reduced
by 90%

Figure 4.15   Solutions to the problem of acid rain

EXPERIMENT

Hypothesis to be tested: Rain is always acid

You will need:
distilled water
pH paper or a pH meter
carbon dioxide in a gas jar (obtained from a cylinder or by allowing an acid to react with a carbonate and collecting the gas – *wear eye protection*)

Method
1 Find the pH of the distilled water.
2 Shake 100 cm³ distilled water with carbon dioxide in a gas jar.
3 Find the pH of the water in the gas jar.

Questions
1 Why is the water shaken with the carbon dioxide?
2 Where does carbon dioxide in the atmosphere come from?

EXPERIMENT

Hypothesis to be tested: Acid rain affects the growth of maize

You will need:
maize seeds
compost
15 flower pots
3 gravel trays
home-made 'acid rain' at pH 3 and pH 4 (made by bubbling sulphur dioxide through distilled water until the required pH is reached in each case)
distilled water brought to pH 5.5 by bubbling carbon dioxide through it (these three solutions should be prepared by a teacher or technician)

Method
1 Plant the seeds in pots full of compost, two seeds to a pot.
2 Label five pots pH 3, five pH 4, and five pH 5.5.
3 Put the pots into gravel trays so that all the pH 3 pots are together, all the pH 4 pots are together and all the pH 5.5 pots are together.
4 Water them with the appropriate 'acid rain' (from above with a watering can).
5 Put them in a sunny position.
6 Water as necessary with the appropriate 'acid rain'. Rotate the position of the trays each week.
7 At weekly intervals after germination, count the number of leaves per pot and calculate the size of the largest leaf in each pot. (You will need to decide how you are going to calculate the size of the largest leaf and how you are going to record the results.)
8 At the end of the experiment, compare the growth of the seedlings by removing the plants from the soil, washing them and then oven-drying them. Compare *dry* masses of the plants.

This experiment will run over a long period of time. In a class experiment the results could be filed on a floppy disc, the students recording their results on to the disc each week. Copies of the results can then be issued to each student at the end of the experiment for analysis.

Questions
1 Why were the trays rotated each week?
2 Why were the seedlings finally dried before weighing?
3 Why was more than one seed planted and watered with each strength of acid rain?
4 It is important to have a population of bacteria and fungi in the soil, so that nutrients are recycled and dead material is broken down. Formulate a hypothesis, and then design an experiment to find out whether acid rain affects these micro-organisms. What results will you record?

─ **Summary** ─

■ Oxides of sulphur and nitrogen are responsible for acid rain.

■ They are produced from the combustion of fossil fuels in power stations, industry and vehicles.

■ Acid rain causes the death of trees, alters the fauna and flora in lakes and can be damaging to human life.

■ Solutions to the problem include
changing the ways in which fuels are burned
cleaning coal and oil before burning
filtering smoke from power stations and factories
filtering smoke from vehicles
conserving energy.

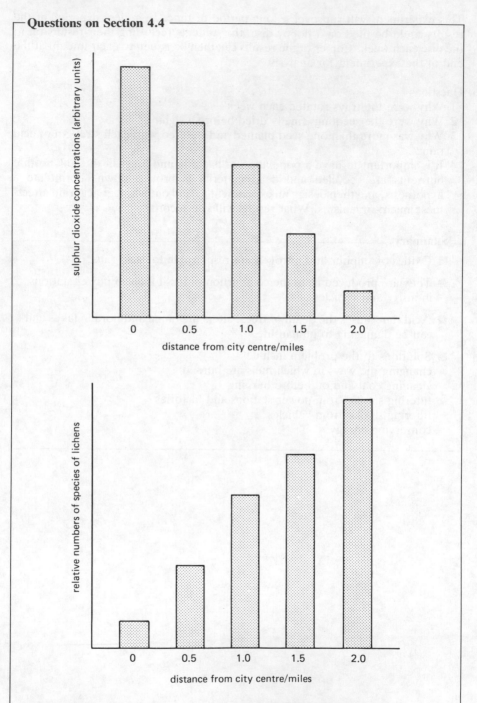

**Figure 4.16   Bar charts representing atmospheric sulphur dioxide
concentrations and numbers of species of lichens
(a type of plant)**

Examine Fig. 4.16.

1 Where does the sulphur dioxide in the air come from?
2 How do levels of sulphur dioxide in the centre of the city compare with those outside the city?
3 Why do the levels change?
4 How might the sulphur dioxide levels change on a windy day?
5 Do you think there is a relationship between the amount of sulphur dioxide in the air and the number of species of lichen?
6 Suggest a hypothesis which could be used to test your answer to question 5.
7 Could you use lichens as a means of finding the sulphur dioxide concentration in the air at the centre of the city? How would you do this? What problems might you have?

⬡5 Food for energy

5.1 Good food – nutrition and energy

You were introduced to the nutrients our bodies need in Sections 1.5 and 3.1. In this section you will find out which foods contain these nutrients, and more about vitamins and minerals.

Protein, carbohydrates and lipids

Look at Fig. 5.1(a), (b) and (c). Note the foods which contains these nutrients.

Exercise

Write down a list of all the foods you ate yesterday.

1 Which foods in your list contained protein?
2 Which foods in your list contained carbohydrates?
3 Which foods in your list contained lipids?
4 Did the food you ate yesterday provide you with all these nutrients?
5 If one was missing, what could you have eaten to supply you with this nutrient?
6 What could happen to you if you do not get this nutrient for a long time? You may need to look at Fig. 5.1 again.

Look at Fig. 5.1(a) and (b) again. Note which people require higher amounts of protein and carbohydrate.

Exercise

1 Why do nursing mothers, babies, children and adolescents have high protein requirements?
2 Explain why nursing mothers, adolescents and adults doing manual work have high carbohydrate requirements.

Minerals and vitamins

These nutrients are required in very small amounts. A varied diet, with plenty of fresh fruits and vegetables, provides an adequate supply of these nutrients.

Look at Table 5.1, which summarises the sources and functions of some of the most important minerals needed by the body.

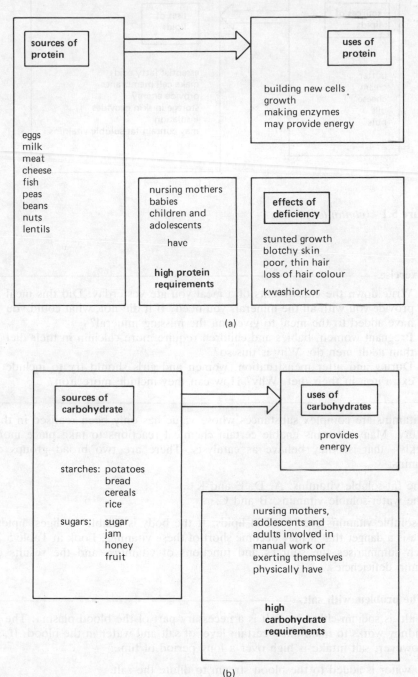

Figure 5.1 (*continued overleaf*)

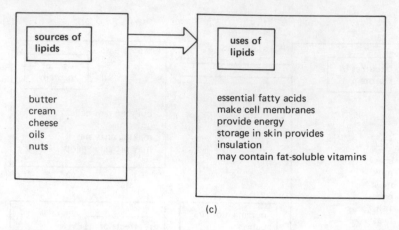

(c)

Figure 5.1 (*continued*)

Exercise

1 Write down the constituents of a meal you ate yesterday. Did this meal provide you with all the minerals you need? If it did not, what could you have added to the meal to give you the missing mineral?

2 Pregnant women, babies and children require more calcium in their diet than adult men do. Why is this so?

3 During and after menstruation, women and girls should try to include extra iron in their diet. Why? How can they include more iron?

Vitamins are complex substances whose value has only been realised in this century. Many vitamins enable certain chemical reactions to take place more quickly – that is, they behave as catalysts. There are two broad groups of vitamins:

- the fat-soluble vitamins: A, D, E and K
- the water-soluble vitamins: B and C.

Fat-soluble vitamins are found in lipids. If the body is unable to digest lipids, there is a danger that it may become short of these vitamins. Look at Table 5.2, which summarises the sources and functions of vitamins, and the results of vitamin deficiencies.

The problem with salt

'Salt' is sodium chloride. Salt is a necessary part of the blood plasma. The kidney works to maintain a certain level of salt and water in the blood. If, however, salt intake is high over a long period of time:

- water is added to the blood stream to dilute the salt
- the volume of blood is therefore increased
- more fluid goes to the tissues

- small vessels carrying blood to the tissues constrict to prevent the tissues being oversupplied
- constriction of small vessels leads to increased pressure in large vessels, which can lead to high blood pressure.

Table 5.1 Sources and functions of minerals

Mineral	Source	Function
calcium	milk, cheese, eggs, dark green vegetables, soybeans, sesame seeds, fish, meat, blackstrap molasses	needed for good bone and teeth structure, muscle contraction and blood clotting
iron	eggs, meat, spinach, dark green vegetables, chick peas, black beans, soybeans, wheat germ, oatmeal, potatoes	forms part of haemoglobin (molecule which combines with oxygen and transports it around the body)
phosphorus	milk, peas, meat, fish, eggs, cottage cheese, almonds, wheat germ, soybeans, pinto and black beans	needed for good bone and teeth structure, and for muscle contraction
potassium	spinach, butter beans, raisins, prunes, oranges, milk, peas, brussels sprouts	takes part in transmission of nervous impulses, and in chemical reactions inside cells
sodium	salt, cheese, eggs, meat, milk, butter, margarine (see the boxed note 'The problem with salt')	part of tissue fluids, including blood, takes part in kidney functioning and transmission of nervous impulses
iodine	water, iodised salt, wide range of foods	forms part of thyroxin (a hormone which controls metabolic rate)
chlorine	as for sodium	contained in gastric juice and tissue fluids, including blood
fluorine	added to some drinking water, tea	needed for strong enamel on teeth, aids the deposition of calcium in bone

Table 5.2   Vitamins

Vitamin	Source	Function	Result of deficiency
A	milk, butter, egg yolk, fish liver oils, liver, carrots, fresh green vegetables	allows vision in dim light, maintains a healthy skin	poor vision in dim light, skin and cornea of eyes become dry, increased susceptibility to disease
B complex	whole cereal grains, wholemeal bread, yeast, liver, egg yolk, peas, beans, fresh green vegetables, milk, nuts, cheese	help to release energy from food, many chemical reactions in cells	functioning of the nervous system and digestive system is disrupted, skin and mucous membranes are affected
	vitamin B_{12} is only found in animal products and in yeast extract	vitamin B_{12} is involved in the development of red blood cells	pernicious anaemia
C	citrus fruits, tomatoes, potatoes, fresh green vegetables, rosehips	involved in making connective tissue, respiration and pigment metabolism	scurvy, bleeding under the skin and in the joints, bleeding gums, poor healing of wounds, irritability, loss of appetite and weight
D	milk, eggs, butter, fish liver oils, sitting in the sun	enables calcium to be absorbed and promotes its deposition in bones and teeth	rickets (softening and deformation of bones)
K	foods containing lipids (bacteria in the bowel can make this vitamin)	involved in the formation of prothrombin in the liver (prothrombin is involved in the clotting of the blood)	bleeding under the skin, blood will not clot

┌───┐
Exercise

As in the previous exercise, consider a meal you ate yesterday. Did this meal provide you with all the vitamins you need? If it did not, what could you have added to the meal to give you the missing vitamin?
└───┘

Energy

Proteins, carbohydrates and lipids can all be used by the body as sources of energy. Table 5.3 shows the energy values of these three nutrients, measured in kilojoules per gram of food.

Look at Table 5.3. Find out which nutrient, gram for gram, can give you the most energy.

The amount of energy a person needs depends on the type of life he or she leads. Look at Table 5.4. Notice that the more active the task, the more energy is required by the body.

Table 5.3 Energy values of some nutrients

Food	Energy value/kJ per gram
carbohydrate (glucose)	16
lipid	37
protein	17

Table 5.4 Approximate energy requirements of various activities

Activity	Energy/kJ per minute
sleeping	5
sitting	6
typing	6
walking	15
walking up and down stairs	40
golf	10–20
carpentry	10–20
driving	10–20
gardening	21–30
dancing	21–30
squash	30+
coal mining	30+
football	30+

Exercise

In a 24-hour period a person has spent:

2 hours walking
½ hour walking up and down
stairs
8 hours typing

1½ hours gardening
4 hours sitting watching television
8 hours sleeping

What is the energy requirement in kilojoules for this person?

EXPERIMENT

Hypothesis to be tested: Fresh orange juice contains more vitamin C than orange squash or carton orange juice

---

You will need:
Petri dish with jelly containing 0.1% dichloro-indophenol
(DCPIP) to a depth of 1 cm – DCPIP colours the jelly blue
cork borer
3 pipettes each 1 cm^3 capacity
pipette filler
1 cm^3 fresh orange juice
1 cm^3 orange squash (diluted as for drinking)
1 cm^3 orange juice from a carton

[DCPIP can be used as a solution instead of in a jelly. The solution turns colourless in the presence of vitamin C. The more vitamin C there is present, the quicker the DCPIP will turn colourless.]

Method
1 Cut three wells in the jelly, using a cork borer.
2 Label the holes A, B and C on the bottom of the dish.
3 Place 1 cm^3 fresh orange juice into well A, 1 cm^3 diluted orange squash in well B and 1 cm^3 orange juice from a carton into well C.
4 Put the top on the Petri dish and leave for several hours. A clear area will develop around each well as a result of the vitamin C in the samples. The larger the clear area, the more vitamin C there is in the sample.
5 Construct a results chart. (You will have to consider what you are going to record as your results.)

Questions
1 How has the vitamin C moved in the jelly?
2 Why was it important to use exactly the same amount of orange sample in each well?
3 Why is a separate pipette used for each sample of juice?
4 Suppose that 1 cm^3 of a solution containing a known quantity of vitamin C was placed into a fourth well in the jelly. How could the diameter of the clear area help you to determine the vitamin C content of the orange samples used in the experiment?

It is believed that cooking vegetables destroys their vitamin C content. Formulate a hypothesis and design an experiment to investigate this belief. Include a list of the apparatus you would require, a method and a results table.

EXPERIMENT

Aim: To compare the energy values of bread and sugar

You will need:
bunsen burner
crucible
copper can
thermometer
tripod
safety spectacles

pipeclay triangle
retort stand, boss and clamp
1 g bread
1 g sugar

Method
Read through the method and then work out what you are going to record. Draw up a results table before you begin the practical work.

1 Put 100 cm³ cold water into the copper can. Support it with the boss and clamp on the retort stand.
2 Record the temperature of the water.
3 Put 1 g bread into the crucible. Wearing eye protection, place the crucible on the pipeclay triangle and heat it until the bread begins to burn. Quickly move the tripod under the copper can. (See Fig. 5.2.)
4 When the food has stopped burning, check that it has completely burnt. If not, heat the food once more with the bunsen. (Make sure you heat it *away* from the copper can. Heat from the bunsen must not reach the can.)
5 When the food has burnt completely, stir the water gently and record its temperature.
6 Repeat the experiment with a fresh lot of cold water, but this time use sugar instead of bread.

Results
By comparing the temperature rises in the water produced by burning the bread and the sugar, you can say which of these two foods has the highest energy value.
   You can work out the exact energy value of the foods as follows.

We know that:

• 100 cm³ cold water has a mass of 100 g, and
• the energy needed to increase the temperature of 1 g water by 1 °C is 4.2×1 joules.

So the energy needed to increase the temperature of 100 g water by 1 °C is 4.2×100 joules.

Your burning food increased the temperature of the water by T °C (T = final temperature of the water − initial temperature). So, the energy used to increase the temperature of 100 g water by T °C must be $T \times 100 \times 4.2$ joules.
   Therefore 1 g food produces $T \times 420$ joules of energy when it is burnt; we say that its energy value is $T \times 420$ joules per gram.

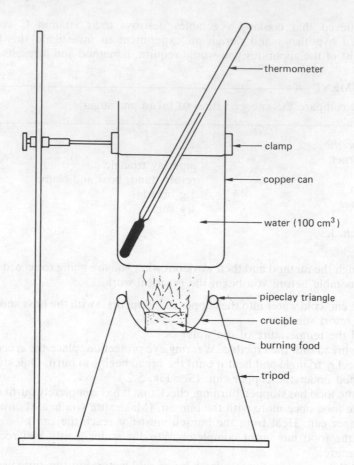

thermometer

clamp

copper can

water (100 cm³)

pipeclay triangle

crucible

burning food

tripod

Figure 5.2   Finding the energy value of a food

Questions

1 Why is it important that heat from the bunsen burner does not reach the copper can?
2 Why is the water stirred before the second temperature is taken?
3 The energy values you found for the sugar and bread may have been much lower than you expected. Why do you think this was so?
4 Look at Fig. 5.3. This apparatus is used to measure the energy value of food more accurately. Study it carefully, and then explain why it gives more accurate results than the apparatus you used in your experiment.

Summary

■ Proteins are found in milk, cheese, eggs, meat, fish, pulses and nuts.

■ Carbohydrates are found in fruit and vegetables and all products containing flour and sugar.

■ Lipids are found in dairy products, oils, animals fats and nuts.

■ The minerals required by the body are calcium, iron, phosphates, potassium, sodium, iodine, chlorine and fluorine.

■ The vitamins required by the body are A, B, C, D and K.

■ Vitamin C can be measured using DCPIP.

■ Carbohydrates, proteins and lipids provide the body with different amounts of energy.

■ The amount of energy a person requires in a day depends on how physically active they are.

■ A bomb calorimeter is used to measure the energy value of food

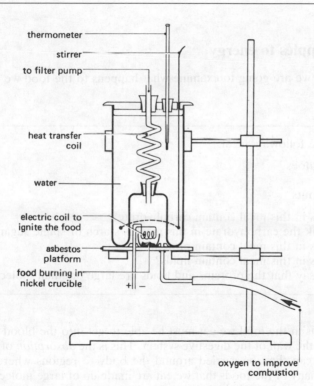

thermometer

stirrer

to filter pump

heat transfer coil

water

electric coil to ignite the food

asbestos platform

food burning in nickel crucible

oxygen to improve combustion

Figure 5.3   A bomb calorimeter

Questions on Section 5.1

1 A person wants to lose weight. He decides not to eat lipids at all.
  (a) Why has he decided not to eat lipids rather than any other type of food?
  (b) Do you think his decision is wise? Explain your answer.
  (c) What advice would you give this person regarding his diet?

2 A newly pregnant woman declares that now she must 'eat for two'.
   (a) Do you agree with this statement?
   (b) What advice would you give on what she should eat?
3 A small child will only eat chips, crisps and sweet biscuits.
   (a) What nutrients are missing from this child's diet?
   (b) What could happen to this child if this remained all he ate?
4 A colleague complains of feeling tired all the time, of not having any
   energy. She also looks very pale. You know that she never eats breakfast
   and only eats small 'fast food' snacks during the day.
   (a) What mineral do you think she is lacking?
   (b) Why does the lack of this mineral give the symptoms described?
   (c) What foods could she eat to provide her with this mineral?

5.2 From apples to energy

In this section we are going to examine what happens to the food we eat.

Exercise

Consider the following meal:

mashed potatoes
chicken leg
brussels sprouts

Which foods in this meal contain carbohydrate?
Do you think the carbohydrate in this meal is starch or single sugars?
Which foods in this meal contain protein?
Which foods in this meal contain lipids?
Would you say that the proteins and lipids are large or small molecules?

The nutrients in the food we eat must be able to get into the blood stream by
passing across the wall of the digestive system. This is the *absorption* of nutrients.
The nutrients can then be carried around the body to regions where they are
needed. But many of the foods that we eat are made up of large molecules – too
large to pass across the wall of the digestive system.

So the large molecules we eat must be broken down into smaller molecules.
This breakdown is called *digestion*. Look at Fig. 5.4, which summarises digestion.
Take note of the end products of digestion. The molecules of these substances are
small enough to pass across the wall of the digestive system.

Special terms are used to describe the way in which our bodies deal with food.
Ingestion is the taking in of food (eating). *Digestion* is the breakdown of large
molecules by enzyme action into smaller ones. *Absorption* is the process by which
these small molecules are taken into the body from the gut, and *egestion* is the
elimination of food that has not been digested.

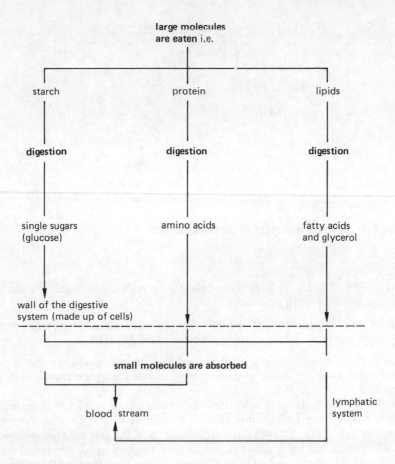

Figure 5.4 Digestion

Teeth

When you eat food or ingest it, you use your teeth and tongue. Do the experiment to investigate the action of your teeth (printed at the end of this section) before continuing.

Fig. 5.5 shows the arrangement of teeth in an adult. Now you should be able to name the different types of teeth in the table you drew up in your experiment. Here is a summary of teeth and their action:

- *incisors*   for biting
- *canines*   for biting harder, tougher food and for tearing
- *premolars* and *molars*   for crushing and grinding (notice the *cusps* or grooves in your own molars – these help the grinding action)

Look at Fig. 5.6(a) and (b). Notice that a small child has only 20 teeth but an adult has 32. Notice also the ages when the particular teeth grow.

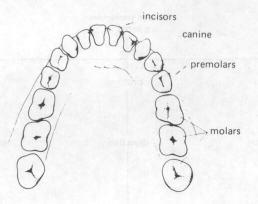

Figure 5.5   The arrangement of teeth in an adult

The child's 20 teeth are sometimes called *baby milk teeth*. Humans start with a smaller number of teeth because

- babies' food is liquid or soft and does not require the hard large teeth of adults
- a baby's jaw bone is too small to accommodate 32 teeth.

The milk teeth need to be replaced as the jaw gets bigger and the diet changes. The stronger adult teeth are of course built to last a lifetime rather than a few years.

The baby milk teeth start to crystallise in the gum six weeks after fertilisation but generally do not appear until the baby is about six months old. As Fig. 5.6 shows, the first teeth to appear are the incisors and the last are the back molars.

As the permanent teeth develop, the roots of the milk teeth are absorbed back into the body. The positions of the milk teeth appear to determine the positions of the permanent teeth, the first of which emerge when the child is about six years old. By the age of 12, all the milk teeth have usually been replaced. The last molars (the wisdom teeth) emerge only after the age of 17 – sometimes not at all.

When you look at your teeth in the mirror you can see only the crowns of the teeth. There are roots in the gums. The number and size of its roots give the tooth stability and strength. Incisors, canines and premolars have one root each, but molars on the upper jaw have three roots and those on the lower jaw have two. Look at Fig. 5.7. Take careful note of all the structures of the tooth.

Tooth decay and gum disease
Tooth enamel is a very hard substance but it can be broken by pressure – for example, if you bite something very hard. It can also be damaged by acids produced in the mouth as a result of bacterial action. These bacteria are very small organisms that live on the surfaces of teeth and in the crevices between them. As the bacteria feed on the food left between the teeth and reproduce, they produce acids which break down the enamel. (But read the boxed note 'Alternative theories of tooth decay'.) Look at Fig. 5.8. Notice which parts of the teeth decay most easily. Notice also how tooth decay can lead to gum disease.

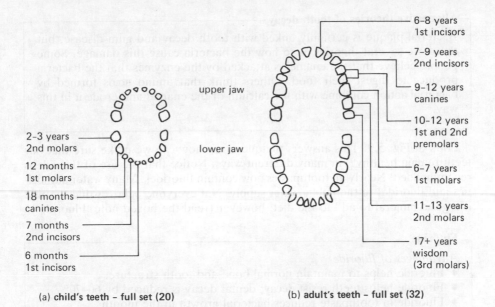

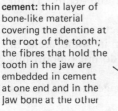

(a) child's teeth – full set (20)

(b) adult's teeth – full set (32)

Figure 5.6 **(a) Child's teeth, (b) adult's teeth, showing the age at which each group of teeth appears**

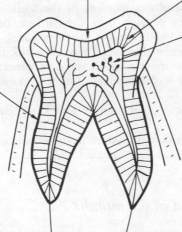

enamel: hard, brittle layer made up of calcium and phosphate arranged as prisms; forms a hard biting surface

dentine: similar to bone in structure, hard but not brittle; strands of living cytoplasm penetrate the dentine from the inside of the tooth

cement: thin layer of bone-like material covering the dentine at the root of the tooth; the fibres that hold the tooth in the jaw are embedded in cement at one end and in the jaw bone at the other

pulp: soft connective tissue containing capillaries and the endings of receptor neurones; the endings are sensitive to changes in temperature and give the sensation of pain; as the tooth matures the channel at the base of the tooth constricts allowing a blood flow insufficient for growth but adequate to maintain life

root: the tooth is held in the jaw bone by collagen fibres, which prevents crushing of blood vessels and nerves when biting occurs

Figure 5.7 Tooth structure

Alternative theories of tooth decay

Bacterial plaque is certainly linked with tooth decay and gum disease, but there are several theories as to how the bacteria cause this damage. Some people believe that the enamel is attacked by the enzymes that the bacteria produce to digest their food. Others think that amino acids formed by bacterial action combine with the calcium of the enamel and erode it in this way.

Look at Fig. 5.9. This answers the question, 'how can we make sure that our teeth remain healthy?' in many different ways. Notice that the use of fluoride is recommended. Nearly all toothpastes now contain fluorides. Many water authorities add fluorides to the public water supply. Not everyone believes that fluoride is a good mineral to add to the diet, however (read the boxed note 'Fluoride').

Fluoride

Advantages of fluoride
- Fluoride helps to maintain normal bone and tooth structure.
- Fluoride helps teeth resist decay; dental decay is reduced by 60–70%.
- Fluoride in toothpaste reduces bacterial growth in the mouth.

The right amount
The optimum concentration of fluoride in water for good tooth structure is 1 part per million.

Not everyone agrees that the fluoridation of water is sensible, however.

Problems with fluoride
- Too much fluoride results in mottling of the teeth. White patches and yellow-brown staining develop over the surface of the teeth. Tooth enamel becomes pitted and corrodes in severe cases.
- Some researchers suggest that there may be a link between cancer and fluoride, and claim there was an increase in cancer in some areas in America where water was fluorinated. Other studies do not support this finding.
- Some researchers suspect that fluoride may cause congenital malformations.
- Cows grazing on fluoride-rich pasture have become crippled.

What happens to food in the mouth?

- Food is chewed, that is physically broken down into smaller pieces by the teeth
- it is moved around the mouth by the tongue
- it is mixed with saliva.

The results of this chewing and mixing process are:

- food is reduced in size, making it easier to swallow

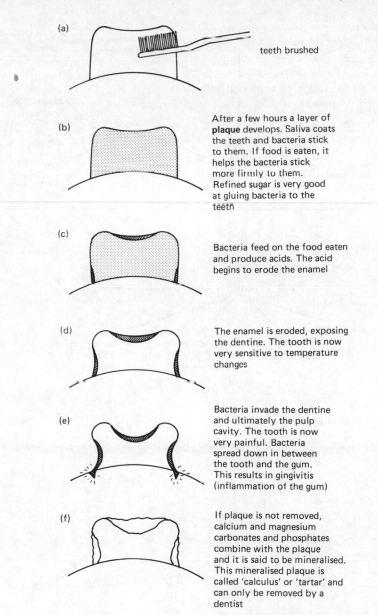

(a) teeth brushed

(b) After a few hours a layer of **plaque** develops. Saliva coats the teeth and bacteria stick to them. If food is eaten, it helps the bacteria stick more firmly to them. Refined sugar is very good at gluing bacteria to the teeth

(c) Bacteria feed on the food eaten and produce acids. The acid begins to erode the enamel

(d) The enamel is eroded, exposing the dentine. The tooth is now very sensitive to temperature changes

(e) Bacteria invade the dentine and ultimately the pulp cavity. The tooth is now very painful. Bacteria spread down in between the tooth and the gum. This results in gingivitis (inflammation of the gum)

(f) If plaque is not removed, calcium and magnesium carbonates and phosphates combine with the plaque and it is said to be mineralised. This mineralised plaque is called 'calculus' or 'tartar' and can only be removed by a dentist

Figure 5.8 Tooth decay and gum disease

- the surface area of the food is increased, thereby exposing a greater surface for enzyme action (read the boxed note 'Surface area and enzyme action)
- food is mixed with saliva; the mucus in the saliva sticks the food particles together and lubricates them, making the food easier to swallow
- the enzyme amylase in the saliva speeds up the breakdown of cooked starch to maltose

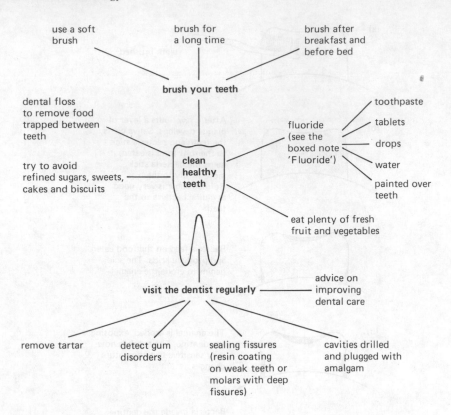

Figure 5.9   Dental hygiene

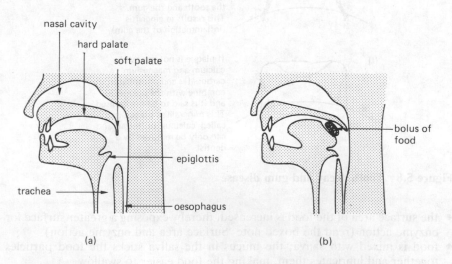

Figure 5.10   (a) Section of the head; (b) the action of swallowing

- water in the saliva softens the food and allows amylase to flow around the food particles
- saliva enables the taste buds on the tongue to be stimulated. This makes these receptor neurones send impulses to the stomach, so that its cells may start to produce digestive juices – in other words, get ready, food on the way!

Surface area and enzyme action

Suppose you have been given a cube of chalk, and asked to grind it up into a powder. Your partner is given the same quantity of chalk but it is already broken up into smaller lumps. Who do you think will have reduced their chalk to a powder first?

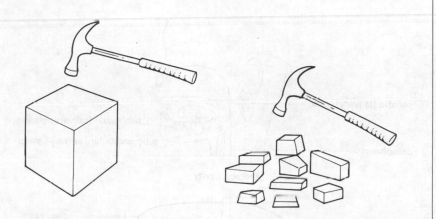

Yes – of course the pile of smaller lumps would be easier to crush. An enzyme experiment would give you similar results. Enzymes speed up the breakdown of molecules on the outside of the particles. The more 'outside' surface there is, the quicker the enzyme action will be.

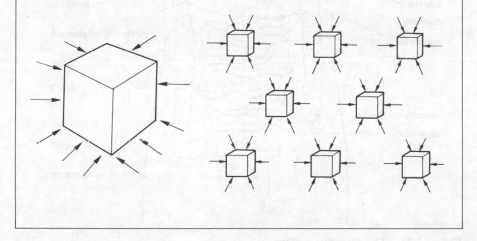

Once the food is thoroughly mixed with saliva and well broken down, the tongue pushes the food against the roof of the mouth (hard palate). Look at Fig. 5.10. Find the hard palate. The food is pushed together into a lump called a *bolus*. The tongue quickly pushes it to the back of the mouth, against the soft palate. A swallowing action follows, passing the bolus into the *oesophagus*. Find the oesophagus in Fig. 5.10. The *epiglottis* automatically covers the entrance to the trachea (the windpipe) when food is being swallowed. This action prevents food 'going down the wrong way'. This is a 'reflex action', an unlearned response – you will find out more about reflex actions in Section 10.2. Find the epiglottis in Fig. 5.10.

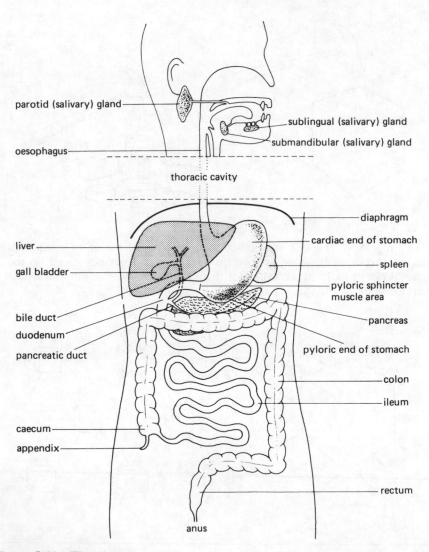

Figure 5.11   The digestive system

Look at Fig. 5.11. This drawing shows the digestive system or *gut*. We will examine what happens to the food as it passes through this system. Find the stomach in the drawing. Put your hand over the part of your body where you think your stomach is. Find the liver in the drawing. Put your hand over the part of your body where you think your liver is. Find the ileum (part of the small intestine) on the drawing and on yourself.

The journey to the stomach

Look at Fig. 5.11 again. Find the *oesophagus*. When food is swallowed it enters the oesophagus. It then quickly passes to the stomach. The oesophagus is not like a drainpipe, it is a closed tube made up of two layers of muscle, one circular and one longitudinal. Look at Fig. 5.12. This shows how alternate contraction and relaxation of these two sets of muscles set up a wave-like ripple along the oesophagus; this pushes the food along it. This action is called *peristalsis*.

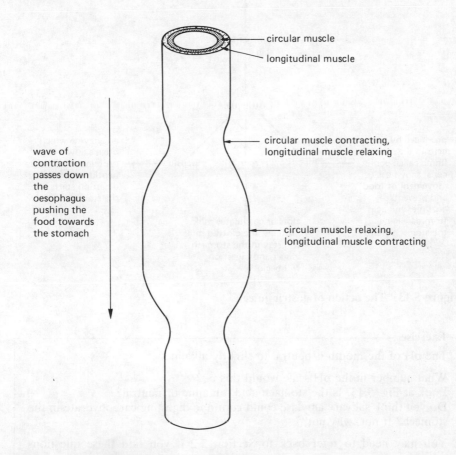

circular muscle
longitudinal muscle

circular muscle contracting,
longitudinal muscle relaxing

wave of contraction passes down the oesophagus pushing the food towards the stomach

circular muscle relaxing,
longitudinal muscle contracting

Figure 5.12   Peristalsis in the oesophagus

Digestion in the stomach

The stomach is a bag with walls made of muscle. Its main functions are to

- store food
- prepare it for digestion in the small intestine
- kill microbes

The wall of the stomach secretes (produces) *gastric juice*. Food is mixed with the gastric juice and digestion begins. Look at Fig. 5.13. This is a summary of what happens to the food in the stomach. Notice that the only food digested is protein. While the food is in the stomach its consistency is changed. A lot of fluid in the form of gastric juice is secreted and there is a lot of mixing by the movements of the muscular walls. The stomach is a very energetic organ! All this movement converts the mixture of food and gastric juice into a runny paste called *chyme*.

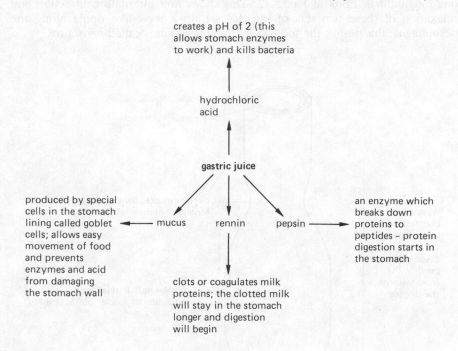

Figure 5.13   The action of gastric juice

Exercise

The pH of the mouth is neutral to slightly alkaline.

What number in the pH scale would this be?
Look at Fig. 5.13. Is the stomach acid, alkaline or neutral?
Do you think salivary amylase could continue digesting carbohydrate in the stomach? If not, why not?

You may need to refer back to Section 2.2 if you find these questions difficult.

Digestion and absorption in the small intestine

Look at Fig. 5.11 again. Find the *duodenum*. Acid chyme enters the duodenum from the stomach. Several secretions are added to the chyme. They come from the *gall bladder* and the *pancreas*.

Find the gall bladder and pancreas in Fig. 5.11. The addition of *pancreatic juice* from the pancreas and *bile* from the gall bladder makes the chyme very watery. This makes enzyme action easier. Look at Figs. 5.14 and 5.15. Notice that there is

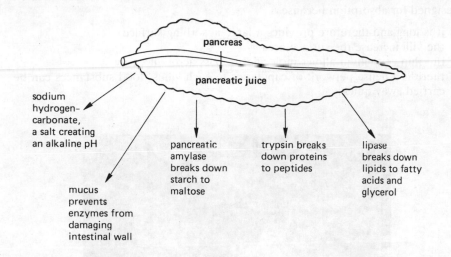

sodium hydrogen-carbonate, a salt creating an alkaline pH

mucus prevents enzymes from damaging intestinal wall

pancreatic amylase breaks down starch to maltose

trypsin breaks down proteins to peptides

lipase breaks down lipids to fatty acids and glycerol

Figure 5.14 The action of pancreatic juice

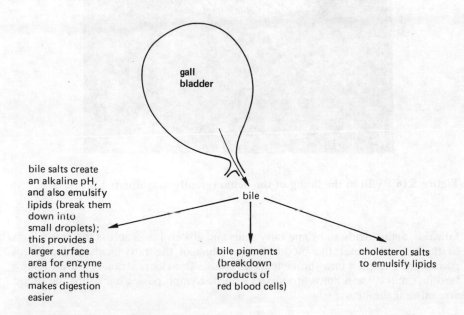

bile salts create an alkaline pH, and also emulsify lipids (break them down into small droplets); this provides a larger surface area for enzyme action and thus makes digestion easier

bile pigments (breakdown products of red blood cells)

cholesterol salts to emulsify lipids

Figure 5.15 The action of bile

another change in pH. Notice also that these secretions digest proteins, starch and lipids.

Fig. 5.11 shows that the duodenum is the first part of the small intestine. The digesting food soon passes into the second part, the *ileum*. Small molecules produced as a result of digestion are absorbed in the ileum. Find the ileum on Fig. 5.11. Then look at Fig. 5.16. This shows the finger-like processes which protrude into the ileum. They are called *villi*. Fig. 5.17 shows a section of a villus. Notice that its surface is rather like velvet, in that it is covered with tiny projections called *microvilli*. Absorption occurs across the walls of the villi. The ileum is well designed for absorption because

- it is long and therefore provides a large absorbing surface
- the villi increase the surface area
- the thin epithelium allows molecules to pass across it easily
- there is a dense network of capillaries to each villus; food substances can be carried away quickly.

Figure 5.16   Villi in the lining of the ileum (greatly magnified)

Glucose, amino acids and some fatty acids and glycerol pass across the epithelium of the villi and enter the blood stream. Most of the fatty acids and glycerol recombine and pass into the lacteals as lipids. The clear, straw-coloured lymph becomes milky when lipids are mixed with it. Lymph passes back into the blood stream near the heart.

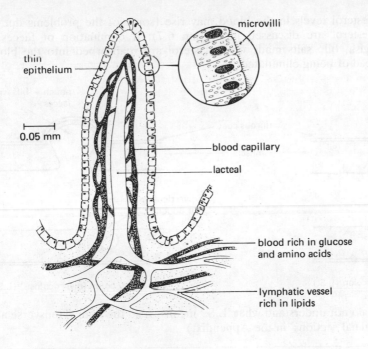

Figure 5.17   Section of a villus (diagrammatic)

The large intestine

Material that enters the colon (the first part of the large intestine) cannot be digested by the body's enzymes. All the digested food has been absorbed. As well as undigested matter (largely cellulose and vegetable fibres – 'roughage'; see the boxed note 'Fibre and a healthy colon'), it contains mucus, dead cells (from the lining of the gut), bile pigments, and a good deal of water. As the material is pushed along the colon by peristaltic action, most of this water is quickly absorbed back into the blood stream. The body would become dehydrated if this reabsorption of water did not take place.

Fibre and a healthy colon

Fibre in our diet is mostly provided by cellulose. It is not digested or absorbed, but it acts as *roughage*. Roughage is important, partly because it helps the intestine to move its contents along, and partly because cereal fibre absorbs water and will keep the faeces moist, soft and bulky. If the diet is low in fibre:

- constipation may become a problem
- the wall of the intestine thickens
- after a long time the surface of the intestine becomes irregular little pockets (*diverticula*) may begin to bulge out of the weak areas (see the diagram below); these 'pockets' may become infected

- cholesterol levels in the blood may rise (some of the problems due to cholesterol are discussed in Section 6.7); if elimination of faeces is sluggish, bile salts made of cholesterol are reabsorbed into the blood instead of being eliminated.

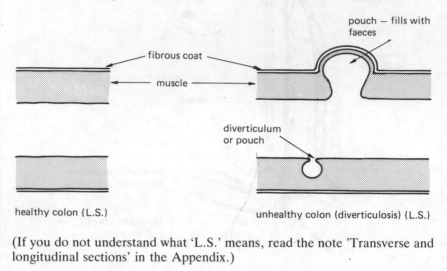

healthy colon (L.S.)   unhealthy colon (diverticulosis) (L.S.)

(If you do not understand what 'L.S.' means, read the note 'Transverse and longitudinal sections' in the Appendix.)

The body does not produce enzymes in the colon but there are many bacteria, which feed on cellulose and other undigested material which are thought to be responsible for releasing vitamin K (essential for the production of a protein necessary for blood clotting, called prothrombin) and some B vitamins. The bacteria living in your colon do not harm you; they help you to stay healthy and we say they are *commensal*.

By the time the undigested material reaches the second part of the large intestine, it is semi-solid. Once the rectum is distended, a nervous impulse produces the urge to expel the faeces.

Assimilation

All the blood vessels in the villi join up to form a large vein, the *hepatic portal vein*. This vein takes the nutrient-rich blood to the liver. Look at Fig. 5.18. This diagram shows what happens to glucose in the blood. Remember that

- the level of glucose in the hepatic vein is lower than that in the hepatic portal vein
- the liver acts as a store for glucose in the form of glycogen; glucose is released from the liver as it is needed
- the liver does many jobs for the body and therefore needs a lot of glucose itself for energy.

Look at Fig. 5.19, which shows what happens to amino acids in the blood. Notice that:

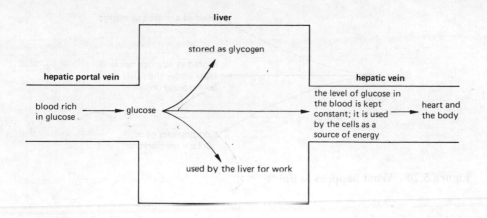

Figure 5.18   What happens to glucose

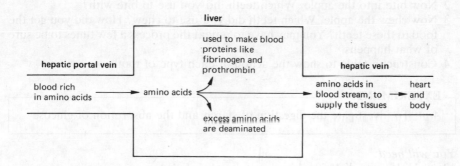

Figure 5.19   What happens to amino acids

- the concentration of amino acids in the hepatic vein is lower than that in the hepatic portal vein
- amino acids cannot be stored; the amino acids that are not required by the body are *deaminated* (that is, the nitrogen portion of the amino acid molecule is removed) and the rest of the molecule is converted to glycogen
- blood proteins are made in the liver.

Look at Fig. 5.20, which shows what happens to lipids.

EXPERIMENT

Aim: To investigate the action of the teeth

You will need:
a mirror   an apple

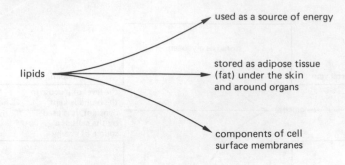

Figure 5.20 What happens to lipids

Method
1 Look at your teeth in the mirror. Notice the different types. How many different types can you see? Make a rough sketch of each type.
2 Now bite into the apple. Which teeth did you use to bite with?
3 Now chew the apple. Which teeth did you use to chew? How did you get the food to these teeth? You may have to repeat the process a few times to be sure of what happens.
4 Construct a table to show the action of each type of tooth.

EXPERIMENT

Aim: To investigate the digestion of starch and the absorption of glucose

You will need:
12 cm Visking or dialysis tubing	starch solution
pasteur pipette	amylase solution
cotton	iodine solution
small beaker	Benedict solution
water bath set at 37 °C	safety goggles

Carry out the experiment shown in Fig. 5.21.

Questions
1 Which food substance was found in the water surrounding the Visking tubing bag?
2 Why was this substance present in the water?
3 What part of the digestive system does the Visking tubing bag best represent?
4 In which ways does the Visking tubing bag act like its counterpart in the digestive system?
5 In which ways does the Visking tubing bag differ from its counterpart in the digestive system?
6 Why was the water bath set at 37 °C?
7 Can you think of a suitable control for this experiment?

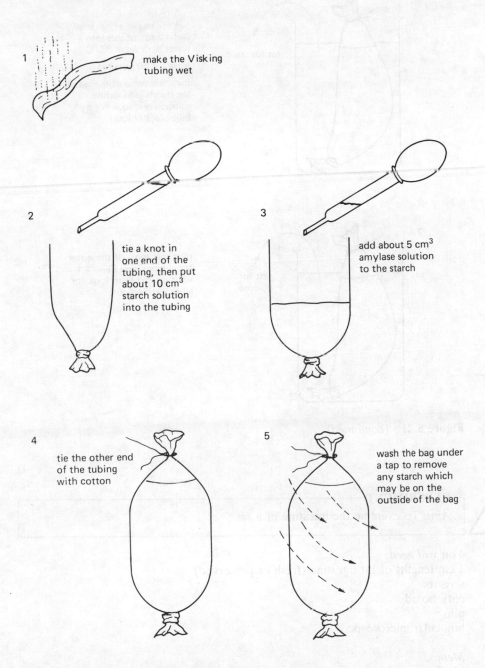

1 make the Visking tubing wet

2 tie a knot in one end of the tubing, then put about 10 cm^3 starch solution into the tubing

3 add about 5 cm^3 amylase solution to the starch

4 tie the other end of the tubing with cotton

5 wash the bag under a tap to remove any starch which may be on the outside of the bag

Figure 5.21 *(continued overleaf)*

6

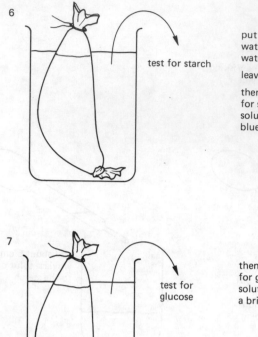

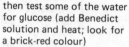

test for starch

put the bag in a beaker of
water and put this into a
water bath set at 37 °C

leave for 30 minutes

then test some of the water
for starch (add iodine
solution and look for a
blue-black colour)

7

test for
glucose

then test some of the water
for glucose (add Benedict
solution and heat; look for
a brick-red colour)

Figure 5.21   (*continued*)

EXPERIMENT

Aim: To examine the intestine of a rat

You will need:
1 cm lengths of rat intestine (fresh or preserved)
scissors
cork board
pins
binocular microscope

Method
1 Cut the intestine along its length.
2 Open it out and pin it securely on to the cork board. The inside surface should
   be facing you.
3 Hold it under a cold tap and wash it thoroughly.

4 Examine with the binocular microscope. Draw what you see as clearly as possible and in the correct proportions.
5 Examine other sections from different regions of the intestine, and compare them.
6 Return all pieces of rat tissue for safe and hygienic disposal. Then wash your hands with soap and water.

Questions
1 What structural features did you observe in the intestine?
2 Rats are weaned at three weeks old (that is, their diet is changed from milk to solid food). Before weaning there is no difference in the structure of the intestine along its length. Why do you think the structure changes after weaning? Formulate a hypothesis as a result of your explanations.

Summary

■ In *digestion* large food molecules are broken down into smaller molecules by *enzymes*.

■ The small molecules are *absorbed* in the small intestine and pass into the blood stream.

■ Food is moved through the digestive system by muscular action – *peristalsis*.

■ Absorbed food molecules are taken to the liver; the liver regulates the amount of glucose and amino acids in the blood.

Questions on Section 5.2

1 A person who has had their gall bladder removed may be advised to limit the amount of lipids they eat. Why is this so?
2 You had a cheese sandwich for lunch.
(a) Explain how and where the sandwich will be digested.
(b) What will be the results of the digestion? (That is, what molecules will be produced?)
(c) Where will the resulting small molecules be absorbed?

5.3 Passing on the energy

The Earth's only incoming source of energy is the Sun. Some of the Sun's energy is trapped by green plants and used to make carbohydrates:

$$\text{energy from the Sun} \xrightarrow{\text{flow of energy}} \text{plant material}$$

Exercise

Look at Fig. 5.22.

What percentage of the Sun's energy is absorbed by green plants?

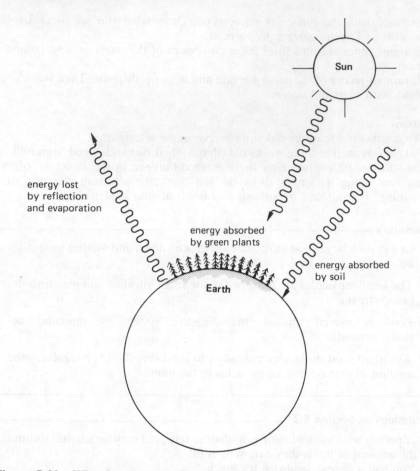

Figure 5.22   What happens to the energy from the Sun: of every 2 000 000 kJ arriving at the Earth's surface, 1 940 000 kJ is lost by reflection, evaporation and absorption by the soil

So only a small amount of the energy reaching the Earth's surface is absorbed by plants and trapped in carbohydrates.

Now look at Fig. 5.23. This drawing shows what happens to the energy in the plant. Notice that the energy stored in the plant is available for other organisms. In Fig. 5.23 the larva is taking some of this energy.

Exercise

The plants in Fig. 5.22 absorbed 60 000 kJ of the Sun's energy; 50 000 kJ of this is stored in new tissue. This is the energy available to other organisms.

What percentage of the energy absorbed by the green plant is stored in new tissue?

What has happened to the energy that has *not* been stored in new tissue?

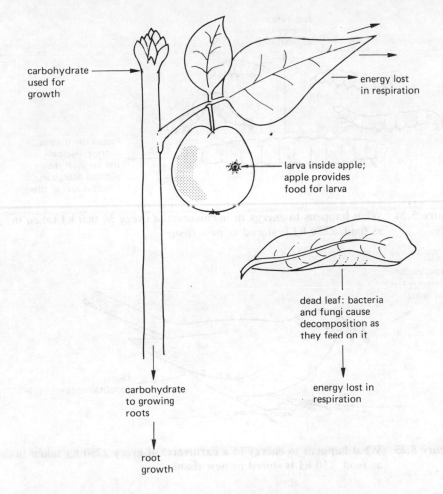

Figure 5.23   What happens to the energy trapped in green plants

The next question to ask is – if larvae and other herbivores eat plant material that has energy equivalent to 50 000 kJ, what happens to this energy? Look at Fig. 5.24.

Exercise

What percentage of the energy taken in by the herbivores is stored as new tissue?

What percentage is lost in respiration, or is unobtainable and lost in faeces?

Which organisms obtain the energy in the faeces?

The energy stored in the new tissue of the herbivore is available to animals that eat herbivores (that is, carnivores). Let's say that birds eat the larvae and take in 2250 kJ of energy. Look at Fig. 5.25 to see what happens to this energy.

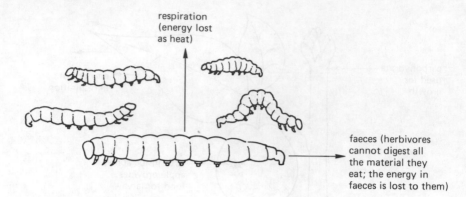

Figure 5.24   What happens to energy in herbivores: of every 50 000 kJ taken in as food, 2250 kJ is stored as new tissue

Figure 5.25   What happens to energy in a carnivore: of every 2250 kJ taken in as food, 150 kJ is stored as new tissue

---

Exercise

What percentage of the energy originally trapped by the green plant is available for carnivores who eat these birds?
If the birds were eaten by cats, what would happen to the energy the cats had taken in?

We can see that there is a relationship between plants, herbivores and carnivores. You can think of this relationship as a chain. Look at Fig. 5.26. This is a *food chain*. Notice green plants are the only organisms that can trap energy and make it available to other organisms. Green plants are called *producers*. All the organisms that consume plants are *consumers*. An animal that eats a green plant is a *primary consumer*. Notice that the primary consumers are herbivores. An animal that eats a primary consumer is a *secondary consumer*. An animal eating a secondary consumer is a *tertiary consumer*. Secondary and tertiary consumers are both carnivores.

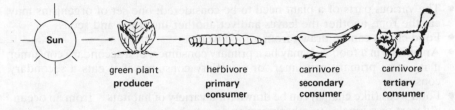

Figure 5.26 A food chain: each arrow denotes a flow of energy

Bacteria and fungi that feed on the dead remains of plants and animals and/or the faeces, are called *detritus feeders* or *decomposers*. Think of the bacteria breaking down the remains of a dead rabbit. Are they producers, primary consumers, secondary consumers or tertiary consumers?

The food chain in Fig. 5.26 represents a relationship between a plant and a few other organisms. But the cabbage, for example, is eaten not only by caterpillars but by rabbits, humans, slugs, snails and other organisms as well. The caterpillars in turn may be eaten by, say, small mammals as well as by birds. A *food web* represents the relationship between a plant and its surroundings better than a simple food chain can. Study the food web in Fig. 5.27. Note the following points:

- Webs begin with green plants, and these plants always get their energy from the Sun
- The arrows (as in food chains) represent the flow of energy

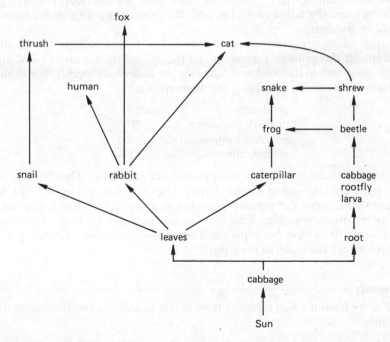

Figure 5.27 A food web

- The various parts of a plant need to be considered: one set of organisms may eat the root, another the leaves and yet another the fruits and seeds
- Food chains can be drawn from a food web
- An animal in a food web may be a primary consumer, or a secondary consumer if it eats a primary consumer, or a tertiary consumer if it eats a secondary consumer
- Food webs (like chains) can be drawn for a variety of habitats – from an ocean, a desert or a jungle, to a back garden or a tree.

The study of food webs can suggest the consequences of removing one animal or plant. For example, removing the rabbits from the food web in Fig. 5.27 would force the fox and the human to eat other available foods. If the only food available to foxes was rabbits and birds, the bird population would fall. In the end the foxes might decrease in numbers. If the numbers of birds fell, the snails would no longer be preyed upon and the cabbage would still be consumed.

Wild populations are generally balanced, that is, there is usually a balance between producers and consumers, and there is enough food for all the organisms. Removal of one type of organism can seriously upset the delicate balance.

The food web in Fig. 5.27 is very simple. You may know more organisms that could be added to this web. The more we know about a particular food web, the more complex it becomes. Diagrams can become so complicated that it is difficult to see the individual links and to grasp the system as a whole. There is therefore a strong case for showing food webs in a more generalised form.

We can start by observing that usually there are far more primary consumers in a food web than tertiary consumers, and that the numbers of secondary consumers generally fall between the two. We can use this knowledge to draw a *pyramid of numbers*.

The procedure for constructing a pyramid of numbers is to count all the producers, all the primary consumers, all the secondary consumers and all the tertiary consumers in the food web, ignoring the individual species. For example, in a study of an area of grassland, the numbers were:

producers	50
primary consumers	20
secondary consumers	5
tertiary consumers	1

These numbers have been drawn as a pyramid in Fig. 5.28(a). The different levels in the pyramids are called *trophic levels*. The pyramid in Fig. 5.28(b) has a different shape: here the primary producer is a single tree which can support many other organisms. Fig. 5.28(c) is an inverted pyramid and represents a human being on whose body parasites are living, on which further parasites (hyperparasites) are living in their turn.

Summary

■ Energy from the Sun is passed from plants to animals and from animal to animal.

■ At each trophic level, energy is lost in respiration and (in animals) in faeces.

■ *Food chains* represent the relationship between plants and animals that feed on each other.

■ *Food webs* represent the relationship between plants and animals in a particular area.

■ Food chains and webs consist of *producers*, *primary consumers*, *secondary consumers* and *tertiary consumers*.

■ The numbers of organisms belonging to different feeding levels can be represented as a *pyramid of numbers*.

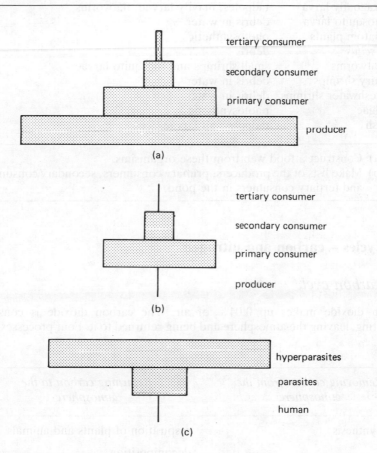

Figure 5.28 Pyramids of numbers

Questions on Section 5.3

1 Some people think that, in view of the world food problem, we should eat more vegetables and little or no meat, since this is a more efficient use of energy. Explain why it is a more efficient use of energy.
2 Draw five food chains. Name the producers, primary consumers, secondary consumers and tertiary consumers.

3 A study was made of a freshwater pond. The fauna and flora present were recorded and observations were made of what the animals ate. The observations are given below.

Animals	Food
tadpoles	mayfly larvae, fairy shrimps, freshwater shrimps
snails	algae
mayfly larvae	diatom plants
dragonfly larva	tadpoles, mayfly larvae, flatworms
mosquito larva	debris in water
diatom plants	photosynthetic
Asellus	debris
flatworms	small shrimps and mosquito larvae
fairy shrimps	debris in water
freshwater shrimps	debris in water
algae	photosynthetic
fish	tadpoles, all larvae

(a) Construct a food web from these organisims.
(b) Make lists of the producers, primary consumers, secondary consumers and tertiary consumers in the pond.

5.4 Cycles – carbon and nitrogen

The carbon cycle

Carbon dioxide makes up 0.04% of air. The carbon dioxide is constantly circulating, leaving the atmosphere and being returned to it. Four processes are at work:

Removing carbon from the atmosphere	Returning carbon to the atmosphere
photosynthesis	respiration of plants and animals
	decomposition
	combustion

■ **Memory check**
What are photosynthesis, respiration and combustion?
Write word equations to represent the processes.
See Sections 4.1, 1.5 and 5.1 respectively.

We can arrange these processes in a *cycle*. We are going to build the carbon cycle up a step at a time – follow the steps carefully.

1 Carbon dioxide is removed from the atmosphere by photosynthesis.

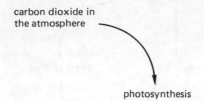

(Formation of carbonate rocks also removes carbon dioxide, but this is a slow process and will not be included in our cycle.)

2 Photosynthesis builds the carbon into new carbohydrates, and ultimately lipids and proteins, in plants. Some of the new food is used in respiration, which will return carbon back into the atmosphere:

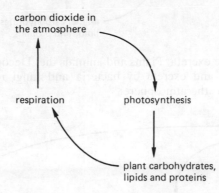

3 Some of the plant carbohydrate, lipid and protein is eaten by animals:

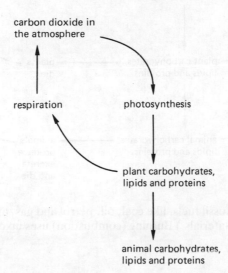

4 Animals also respire, thereby returning carbon (as carbon dioxide) back into the atmosphere:

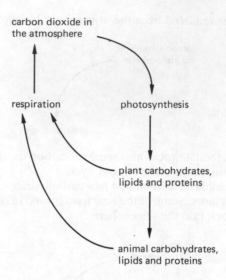

5 Animals produce excreta. Plants and animals die. Decomposition (rotting) of dead organisms and excreta by bacteria and fungi releases more carbon dioxide back into the atmosphere:

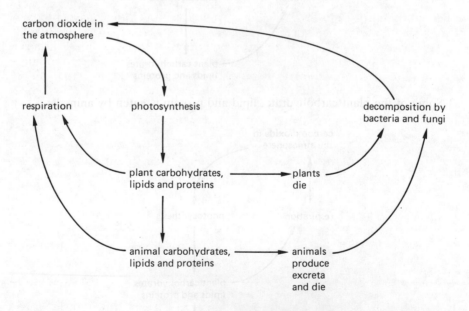

6 Peat, wood and fossil fuels, like coal, oil, petrol and gas, are used for fuel. (All these are plant materials.) Burning (combustion) uses up oxygen and produces carbon dioxide:

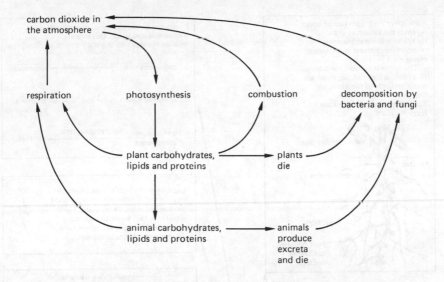

This diagram is now a completed carbon cycle. Spend a few minutes going over the stages in the cycle. Make sure that you understand how the cycle goes on and on. It has no beginning and no end.

The nitrogen cycle

The element nitrogen makes up 80% of the atmosphere. It is found in all proteins. Like carbon, nitrogen is both removed from the atmosphere and returned to it.

Look at Fig. 5.29, which summarises the processes involved in the nitrogen cycle. Notice that nitrogen is converted into nitrates. Plants cannot use atmospheric nitrogen to make protein, but they can use nitrates. Nitrates are therefore the link between atmospheric nitrogen and plants – and, of course, the animals that eat the plants. Let us organise the processes in Fig. 5.29 into those that remove nitrogen from the atmosphere and those that replace it:

Removing nitrogen from the atmosphere	Returning nitrogen to the atmosphere
bacteria in nodules of leguminous plants	denitrifying bacteria
free-living bacteria in the soil	
lightning	

Decomposition, mentioned in Fig. 5.29, has been left out of these lists. This is because decomposition makes nitrates available to plants again, but it does not remove or replace nitrogen. Let's look at those processes that remove nitrates from the soil and those that replace nitrates:

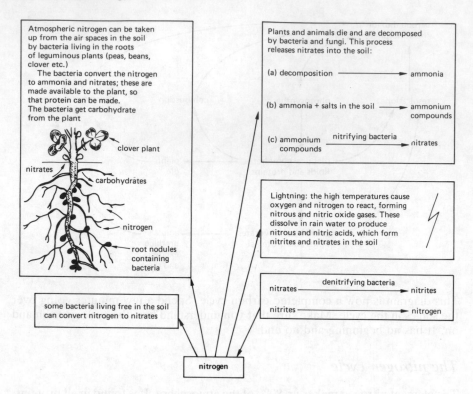

Figure 5.29   Processes in the nitrogen cycle

Removing nitrates from the soil	Returning nitrates to the soil
plants making protein	free-living bacteria in the soil
denitrifying bacteria	bacteria in nodules of leguminous plants
	lightning
	nitrifying bacteria

Now we are going to build up the nitrogen cycle. Follow the steps very carefully.

1 Atmospheric nitrogen is removed from the atmosphere.

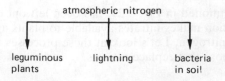

2 Nitrates are formed in these three processes. These enable plants to make protein:

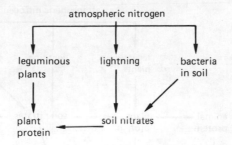

3 Plant protein can become animal protein if the plant is eaten. Or it can become dead plant material:

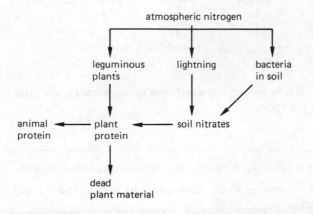

4 Animal protein can be converted into the animal's excretory products. Or it can become dead animal material:

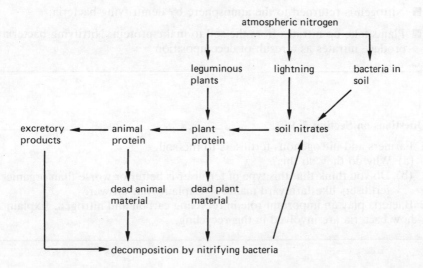

5 Soil nitrates are removed by denitrifying bacteria:

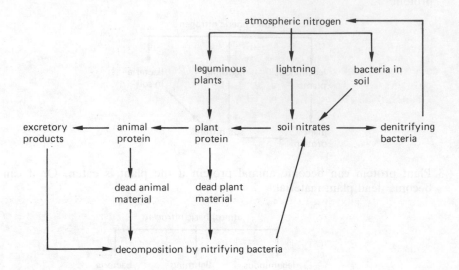

This diagram is now a completed nitrogen cycle. Spend a few minutes going over the stages in the cycle.

Summary

- Carbon is removed from the atmosphere by photosynthesis.

- Carbon is returned to the atmosphere by decomposition, respiration and combustion.

- Nitrogen is removed from the atmosphere by leguminous plants, lightning and bacteria in the soil, and is converted to nitrates.

- Nitrogen is returned to the atmosphere by denitrifying bacteria.

- Plants take up nitrates from the soil to make protein. Nitrifying bacteria produce nitrates as a result of decomposition.

Questions on Section 5.4

1 Farmers add nitrogenous fertilisers to the soil.
  (a) Why do they do this?
  (b) Do you think that this type of fertiliser is better or worse than organic fertilisers like farmyard manure? Explain your answer.
2 Bacteria play an important role in recycling carbon and nitrogen. Explain how bacteria are involved in this recycling.

5.5 **Energy and human use**

It is likely that you will use more energy in the next five years. Consider also that the Earth's population is rising. The more people there are the greater will be the demand for energy. Look at Table 5.5. Notice the huge rise in energy demand expect by 2070.

Table 5.5 Projected energy demands

Year	*Energy requirement*
1975	7.3 TW (1 TW = 1 terawatt
2000	18.1 TW = 10^{12} W)
2070	68.6 TW

The estimates are based on expected population growth and increasing demands from developing countries.

Energy can be supplied by renewable and non-renewable sources. A *renewable source* is one which is not used up; there will always be a supply of this form of energy. But the energy stored in fossil fuels (coal, oil and natural gas) or in the radioactive substances used in nuclear power stations cannot be replaced once it has been used. These are *non-renewable sources*.

Look at Fig. 5.30. This represents the sources of energy for the United Kingdom. Of these, only hydroelectricity is a renewable energy source, and that makes up only 1% of the whole. So we depend on non-renewable energy sources. But how long will these sources last? If it is used at a rate of 70 TW a year our coal will last for 106 years. Oil and gas reserves are much lower than those of coal. Fossil fuels are used as raw materials to make things like fabrics, plastics and drugs, as well as for fuels. And we know that burning fossil fuels adds to the greenhouse effect (Section 4.3). It is wise then, to seek alternatives. Many people think that nuclear power is the most obvious alternative. Let us consider this form of energy.

There are two types of nuclear reaction that release energy: nuclear fission and nuclear fusion. All the nuclear power stations that have been built so far depend on nuclear fission. A great deal of effort has been put into research on nuclear

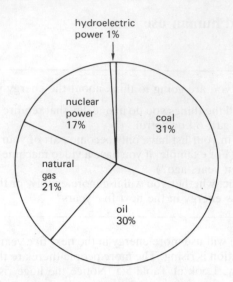

Figure 5.30   Sources of energy in the United Kingdom

fusion reactors, but the problems have proved very great and no commercial fusion reactors have yet been built.

Look at Table 5.6. Notice that the fission fuels cannot be expected to supply energy for the Earth indefinitely. Fusion reactors look much more promising. But designing safe and efficient fusion reactors may prove to be technically impossible. And we cannot ignore the dangers from radiation and the enormous costs of building these reactors and keeping them safe. Nuclear reactors also produce waste that can remain hazardous for hundreds of years. Disposing of this waste supply is an increasing problem.

Table 5.6   Energy from nuclear fuels

Fuel	Reactor	Amount of world energy requirement that could be supplied	Number of years fuel would last
Fission:			
uranium-235	thermal reactor	10%	40–80
uranium-238	fast breeder reactor	all	170
Fusion:			
deuterium		all	10 000
lithium and deuterium		all	tens of thousands

— Exercise ——————————————————————

Do you think nuclear energy is an obvious alternative to fossil fuels? Give reasons for your answer.

Clearly alternatives to nuclear power must be investigated. It is sensible to consider renewable energy sources. Look at Table 5.7. Notice that energy from these sources can be stored in chemical form or used to make electricity.

Table 5.7   Renewable energy sources

Renewable energy source	Method of converting the energy to a usable form	Energy available
solar energy	Photoelectric cells trap the Sun's energy	electricity
solar energy	Mirrors focus light on to a solar furnace. The heat makes steam which drives a turbine. (Small solar furnaces can be used for heating and cooking)	electricity
solar energy	Using sunlight to split water into oxygen and hydrogen. Hydrogen can be used as fuel or in the chemical industry	hydrogen
solar energy	Using organic plant material. Wood is used directly. Methane and methanol can be obtained from the fermentation of plant material. These can be used as fuels or chemical raw materials	methane, methanol
ocean thermal energy	The Sun warms the sea. Warm sea water is passed over pipes that carry the heat energy to a turbine. Large floating ocean thermal plants are proposed for tropical seas	electricity
hydroelectric energy	A trapped body of water is used to turn turbines	electricity
wind energy	Used to turn turbines	electricity
wave energy	Energy from the wind is converted into energy of ocean waves. This energy can be used to turn turbines	electricity

Table 5.7 (*continued*)

Renewable energy source	Method of converting the energy to a usable form	Energy available
tidal energy	Sea water is trapped at high tide. At low tide it is allowed to flow through hydroelectric turbines	electricity
geothermal energy	Activity within the Earth's crust produces heat energy. This may be harnessed where there are hot springs or geysers.	hot water or steam
	In some areas cold water is passed through a pipe down to hot dry rocks deep underground. The heated water or steam is returned to the surface by a second pipe	hot water or steam

Summary

■ Energy sources for humans are non-renewable or renewable.

■ *Non-renewable sources* include coal, oil, natural gas and radioactive substances.

■ *Renewable sources* include sunlight, fuels from plant material, ocean thermal energy, hydroelectric energy, wind energy, wave energy, tidal energy and geothermal energy.

Questions on Section 5.5

1 Processes that trap solar energy using photoelectric cells or mirrors are increasing in efficiency.
   (a) What are the main problems with using sunlight as a source of energy?
   (b) Can you think of any ways in which these problems could be overcome?
2 Wind power can be a useful source of energy in some areas. Why is it not used on a large scale?
3 What environmental problems can be created by using hydroelectric power?
4 Hydroelectricity, wind energy and wave energy are indirect forms of solar energy. Explain why this is so.

PROJECT WORK

1 What might be the consequences of an accident at a nuclear power station? Find out what nuclear power station accidents have occurred and what happened after the accidents.
2 Find out what the United Kingdom does with its nuclear waste. Compare this with what happens in other countries.
3 Gather information on alternative energy sources. Consider their values on a small scale and on a large scale. Find out what the problems are in harnessing renewable energy sources.

6 Transport

6.1 Blood

You know what blood looks like. You probably know too that blood moves around the body. As it does so it acts as a transport system, picking up and delivering the following:

- oxygen from the lungs to the tissues
- carbon dioxide from the tissues to the lungs
- food from the gut to the tissues
- waste products, like urea, from the liver to the kidneys
- heat energy from the liver and muscles to cooler parts of the body
- chemical messengers (hormones) from the glands to other 'target organs'.

If a tube of blood is spun in a centrifuge, it will look like the one in Fig. 6.1. Notice that the blood is made up of *cells* and *plasma*. Notice too that there are two types of cell: *red cells* and *white cells*. There are more red cells than white ones.

Spinning in a centrifuge separates out the denser parts of the blood. You can see from Fig. 6.1 that red cells are the densest. The least dense component of blood is the straw-coloured liquid called plasma. White blood cells have a density midway between those of plasma and red cells.

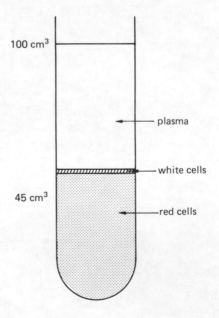

Figure 6.1   Blood after centrifuging

Let's take a closer look at plasma. It contains

- water
- salts
- glucose
- amino acids
- plasma proteins (prothrombin and fibrinogen)
- dissolved gases: oxygen, carbon dioxide (as hydrogencarbonate ions) and nitrogen
- hormones
- urea

The amount of each substance in the plasma varies slightly, depending on where the blood is in the body. For example, in Section 5.2 you saw that glucose is stored in the liver. After a meal, blood going to the liver from the gut contains more glucose than blood leaving the liver.

Now let's consider the red cells. Look at Fig. 6.2. Find the red cells. Notice that a red blood cell has no nucleus. It is shaped like a disc that is concave on both sides (biconcave). Red blood cells are very small – there are 5 million in every cm^3 of blood. Fig. 6.3 is a life history of red blood cells. You should study this carefully.

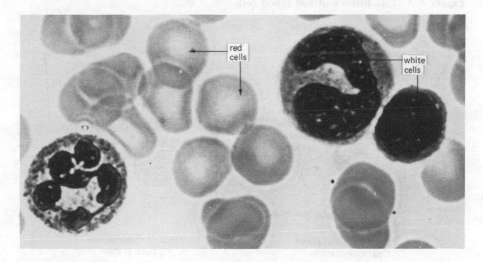

Figure 6.2   Blood cells

Find the white blood cells in Fig. 6.2. Notice that two kinds of white blood cell are labelled. Both have nuclei, but only one has a regular shape.

Fig. 6.4 gives the life histories of these white cells. Study this carefully. Look at Fig. 6.2 once more. Can you name the types of white blood cell shown there?

Blood *platelets* are another type of blood cell. They are very small compared to the other cells. Fig. 6.5 gives the life history of blood platelets. Study this carefully.

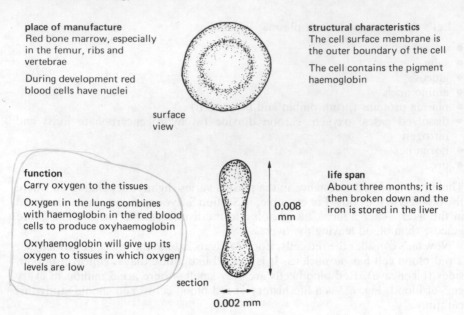

place of manufacture
Red bone marrow, especially in the femur, ribs and vertebrae

During development red blood cells have nuclei

surface view

structural characteristics
The cell surface membrane is the outer boundary of the cell

The cell contains the pigment haemoglobin

function
Carry oxygen to the tissues

Oxygen in the lungs combines with haemoglobin in the red blood cells to produce oxyhaemoglobin

Oxyhaemoglobin will give up its oxygen to tissues in which oxygen levels are low

section

0.008 mm

0.002 mm

life span
About three months; it is then broken down and the iron is stored in the liver

Figure 6.3   Life history of red blood cells

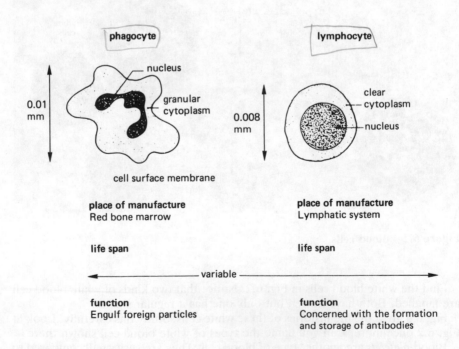

phagocyte

nucleus

granular cytoplasm

0.01 mm

cell surface membrane

place of manufacture
Red bone marrow

life span

function
Engulf foreign particles

lymphocyte

clear cytoplasm

nucleus

0.008 mm

place of manufacture
Lymphatic system

life span

variable

function
Concerned with the formation and storage of antibodies

Figure 6.4   Life histories of white blood cells

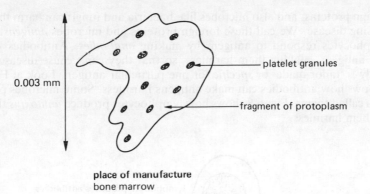

0.003 mm

platelet granules

fragment of protoplasm

place of manufacture
bone marrow

life span
unknown

function
releases a substance that causes blood to clot

Figure 6.5 Life history of blood platelets

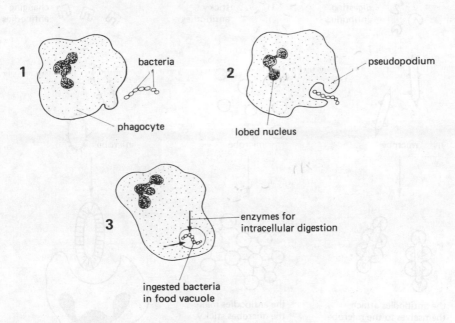

1

bacteria

phagocyte

2

pseudopodium

lobed nucleus

3

enzymes for
intracellular digestion

ingested bacteria
in food vacuole

Figure 6.6 A phagocyte engulfing bacteria

You can see from Figs. 6.4 and 6.5 that white blood cells and platelets help to protect the body. How do these cells fulfil their function?

Look at Fig. 6.6. Notice that phagocytes can change shape and engulf foreign particles. they can also move from capillaries to infected areas of the body.

Foreign proteins, and also microbes like bacteria and fungi, can harm the body by causing disease. We call these foreign proteins and microbes *antigens*.

Lymphocytes respond to antigens by making *antibodies*. Antibodies will act against antigens, making them harmless, so that they can't cause disease. Each antibody is 'tailor-made' or *specific* for one particular antigen. Look at Fig. 6.7. This shows how antibodies can make antigens harmless. Some microbes produce poisons called toxins. Fig. 6.8 shows how lymphocytes produce *antitoxins* that will make them harmless.

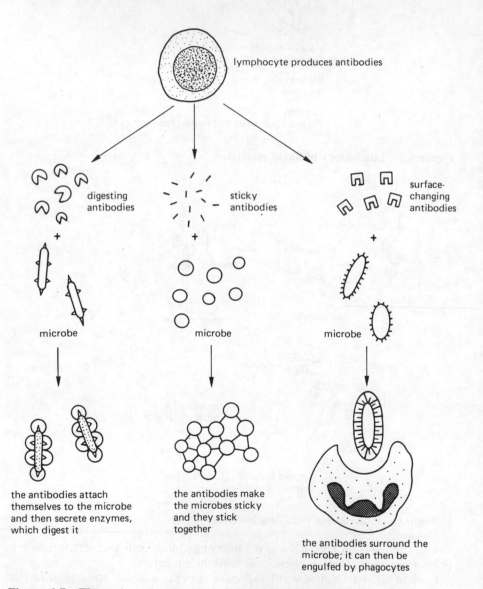

Figure 6.7   The action of antibodies produced by lymphocytes (not drawn to scale)

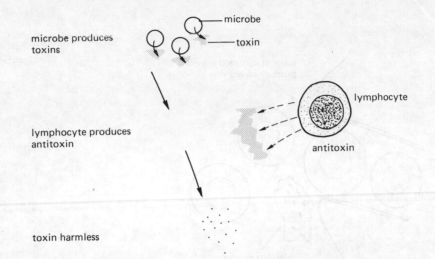

Figure 6.8 The action of antitoxins

The clotting of blood is important because it prevents blood loss and also prevents microbes from entering the body. Fig. 6.9 describes the process of blood clotting.

■ **Memory check**

Prothrombin and fibrinogen are plasma proteins.
Do you remember where they are made?
Which vitamin is necessary to make prothrombin?
See Sections 5.1 and 5.2.

┌─ **Summary** ───

■ Blood is made up of red and white cells, platelets and plasma.

■ Red cells contain haemoglobin and carry oxygen around the body.

■ White cells protect the body by engulfing foreign particles, and also producing antibodies and antitoxins.

■ Platelets are involved in blood clotting.

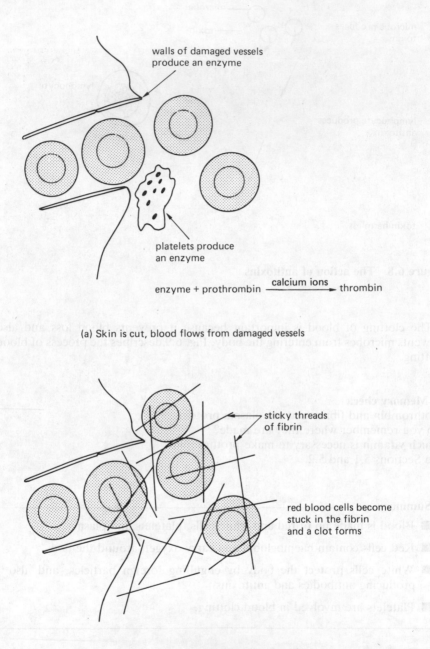

walls of damaged vessels
produce an enzyme

platelets produce
an enzyme

$$\text{enzyme} + \text{prothrombin} \xrightarrow{\text{calcium ions}} \text{thrombin}$$

(a) Skin is cut, blood flows from damaged vessels

sticky threads
of fibrin

red blood cells become
stuck in the fibrin
and a clot forms

(b) thrombin + fibrinogen → sticky threads of fibrin

Figure 6.9   The clotting of blood (not drawn to scale)

Questions on Section 6.1

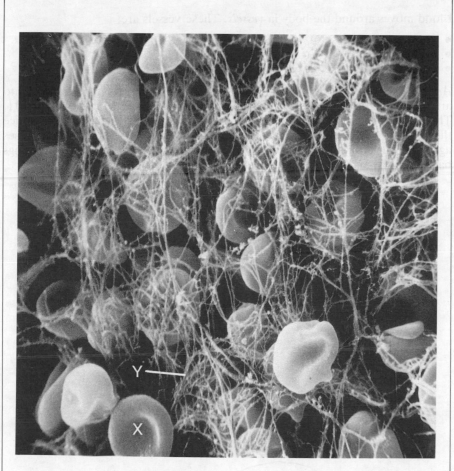

Figure 6.10

1 Look at Fig. 6.10.
   (a) What is X?
   (b) What is Y?
   (c) What has caused Y to form?
   (d) Why is it important that Y has formed?
2 Your friend has had a blood test and her haemoglobin level has been found to be low.
   (a) What is haemoglobin?
   (b) Why is haemoglobin important?
   (c) How might your friend be feeling if her hacmoglobin level is low?
   (d) What foods could she eat to help to increase her haemoglobin levels?

6.2 **Blood vessels**

Blood moves around the body in *vessels*. These vessels are:

● arteries
● capillaries
● veins

Arteries carry blood away from the heart. The heart contracts to push blood into arteries. Therefore the pressure of the blood in arteries is high. Look at Fig. 6.11. This diagram and photograph show the structure of the wall of an artery. Notice the thick muscle and elastic layer. As blood is forced into an artery is presses against the artery walls, stretching the elastic tissue. The elastic tissue then springs back inwards, and the blood is squeezed onwards along the artery.

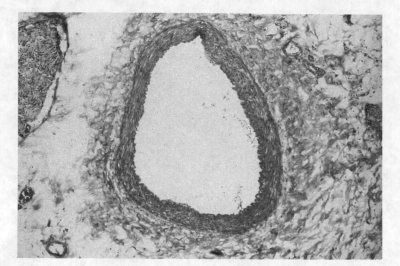

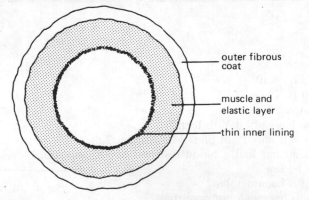

Figure 6.11   Structure of an artery (look at the note 'Transverse and longitudinal sections' in the Appendix)

Arteries divide into smaller vessels called *arterioles*. The muscle in the wall of these vessels helps to regulate blood flow to the tissues of the body. Arterioles divide further to give rise to very small vessels called *capillaries*. Capillaries form a network of vessels called a *capillary bed* in the organs of the body. The capillary bed enables each living cell to be near a flow of blood.

Capillaries are so small that red blood cells have to squeeze through them. Look at Fig. 6.12. Notice that the walls of the capillaries are only one cell thick.

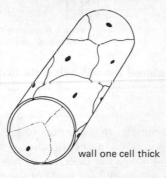

wall one cell thick

Figure 6.12 Structure of a capillary

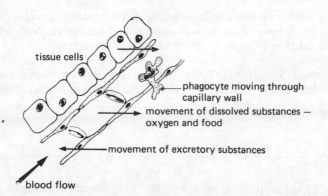

tissue cells

phagocyte moving through capillary wall

movement of dissolved substances — oxygen and food

movement of excretory substances

blood flow

Figure 6.13 Relationship between cells of the body and capillaries

Now look at Fig. 6.13. This drawing represents the relationship between tissue cells and capillaries. Notice that

- nutrients, like glucose and amino acids, and also oxygen and hormones can diffuse from the capillary into the tissue fluid and thence into the cells
- phagocytes can squeeze through capillary walls and move to areas of infection
- waste can diffuse from the cells into the tissue fluid and then into the capillary.

Capillaries leave the organs and join up to form small vessels called *venules*. Venules join up to form *veins*. Blood flows back to the heart in veins. Blood in veins is under low pressure because it is a long way from the heart and it has lost fluid and pressure as it travelled through the capillary bed. Blood flow in veins is maintained by

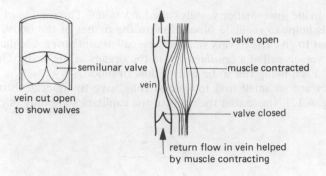

Figure 6.14   Movement of blood through veins: the action of valves

- large veins squeezed by muscles that are next to them (look at Fig. 6.14)
- valves, ensuring that blood flows in one direction – back to the heart; look at Fig. 6.14 once more, and notice the direction of blood flow
- gravity, although this is only useful for parts of the body above the heart
- breathing in; movements of the rib cage and diaphragm encourage blood to be sucked along the veins and towards the heart.

Look at Fig. 6.15, and compare it with Fig. 6.11. Notice that veins have thinner walls than arteries have; there is less muscle and elastic tissue. Notice too that the space for blood flow is larger in the vein than in the artery.

The arrangement of arteries, veins and capillaries is summarised in Fig. 6.16 (page 156). Remember that every organ in the body has its own capillary bed.

Summary

Arteries	*Veins*
■ Carry blood away from the heart	■ Carry blood to the heart
■ Apart from the pulmonary artery they carry oxygenated blood	■ Apart from the pulmonary vein they carry deoxygenated blood
■ Blood flows in pulses under high pressure	■ Blood flows smoothly under low pressure
■ Vessels have a thick muscle and elastic layer	■ Vessels have a thin muscle and elastic layer
■ No valves	■ Valves present
■ Small space for blood to flow through	■ Large space for blood to flow through

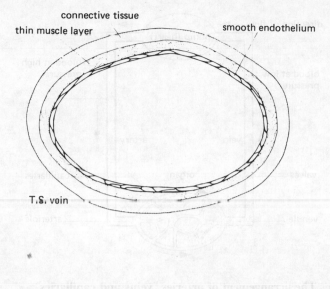

connective tissue
thin muscle layer   smooth endothelium

T.S. voin

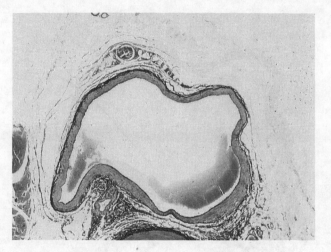

Figure 6.15   Structure of a vein

Questions on Section 6.2

Look at Fig. 6.17.

1 Which of these vessels is a vein? Explain your answer.
2 Draw a longitudinal section of a vein.
3 Draw a longitudinal section of an artery.
4 How is blood moved through veins?

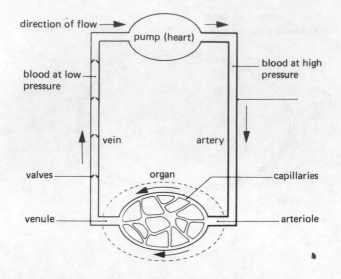

Figure 6.16   **The arrangement of arteries, veins and capillaries**

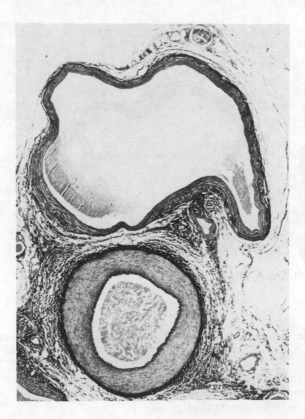

Figure 6.17

6.3 **The heart**

The heart lies in the chest cavity, next to the lungs; all these organs are protected by the rib cage. The heart is a muscular pump, responsible for keeping the blood moving around the body.

Look carefully at Fig. 6.18. The heart is made up of four chambers. The *atria* are the two top chambers and have thin walls. The atria receive blood from the veins. The *ventricles* are below the atria and have thicker walls. The ventricles push blood through the arteries. Find the atria and ventricles in Fig. 6.19.

Four valves ensure that blood can only flow in one direction through the heart. Find the four valves in Fig. 6.18:

- tricuspid valve
- bicuspid valve
- pulmonary valve
- aortic valve

The pulmonary and aortic valves are called the *semilunar* valves. The name comes from the half-moon shapes of these valves. Notice that the tri- and bi-cuspid valves are secured by strong tendons called *chordae tendinae*.

The heart in action

We are going to follow a red blood cell through the heart. Follow this passage on Fig. 6.18.

1 Deoxygenated blood enters the *right atrium* via the *vena cava*. Find the right atrium and the vena cava.
2 When the atrium is full it contracts, pushing blood through the *tricuspid valve* and into the *right ventricle*. Find the tricuspid valve and the right ventricle.
3 When the ventricle is full it contracts, pushing blood up into the *pulmonary artery*. The force of the blood pushes the tricuspid valve closed, preventing blood from flowing back into the right atrium. Find the pulmonary artery. Look at Fig. 6.19(a) and (b), which show how the blood is passed through the right side of the heart.
4 Once the right ventricle stops contracting and starts to relax once more, the pulmonary valve closes, preventing blood from flowing back into the right ventricle. Find the pulmonary valve.
5 The deoxygenated blood goes to the lungs to become oxygenated.
6 Oxygenated blood returns to the *left atrium* via the *pulmonary vein*. Note that there are two veins from the right lung and two veins from the left lung. Find the pulmonary veins and the left atrium.
7 When the atrium is full it contracts, pushing the blood into the *left ventricle*. Find the left ventricle.

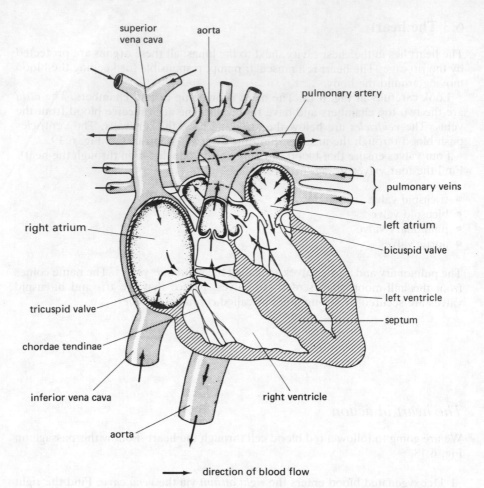

direction of blood flow

Figure 6.18   Structure of the heart. The pulmonary valve (a semilunar valve) is at the opening of the pulmonary artery. The aortic valve is at the opening of the aorta

8 When the left ventricle is full it contracts, pushing blood into the *aorta*. The pressure forces the *bicuspid valve* closed, preventing blood from flowing back into the left atrium. Find the aorta and bicuspid valve.

9 The *aortic valve* prevents blood from flowing back into the left ventricle once contraction ends. Find the aortic valve.

We have followed a red blood cell through the heart but when the heart is in action, one cavity does not work at a time. Instead:

● both atria fill with blood at the same time
● both atria contract at the same time
● so both ventricles fill with blood when the atria contract
● both ventricles contract at the same time.

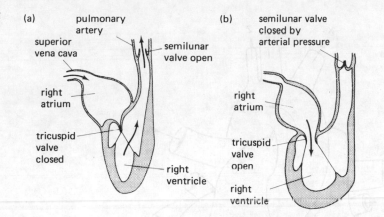

Figure 6.19   Action of the heart (right side only): (a) the ventricle contracts, and blood is sent to the lungs via the pulmonary artery; (b) the atrium contracts, pushing blood into the ventricle

At the end of this sequence the heart relaxes while the atria fill with blood once more. Each repeating sequence is a *heart cycle* or a *heartbeat*. The heart beats about 76 times a minute. The number of beats per minute is the *heart rate*. The heart rate varies from person to person and increases during exercise, so that oxygenated blood and glucose can be carried to the tissues more quickly. The actual heart rate for a person depends on age, health and exercise.

You can find your heart rate by counting your *pulse*. You can feel your pulse where an artery crosses a bone or some hard tissue – at your wrist, for example. The sudden expansion and elastic recoil of an artery (described under 'Arteries' in Section 6.2) causes a ripple to travel along its walls away from the heart. This ripple causes the pulse.

Exercise

Look at Fig. 6.20. Find the atria and ventricles. Then find the pulmonary artery.

1 Where are these vessels taking the blood?
2 Why are there two pulmonary arteries?

Notice that the blood returns to the heart via four pulmonary veins. Now find the aorta.

The first artery to arise from the aorta is the *coronary artery*. You can see the coronary artery in Fig. 6.20 but you cannot see where it arises from the aorta. This vessel is supplying the heart.

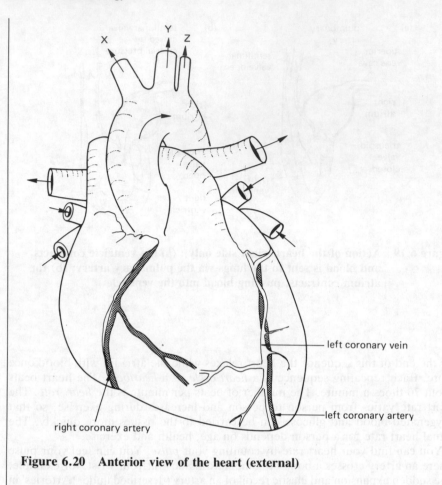

left coronary vein

left coronary artery

right coronary artery

Figure 6.20   Anterior view of the heart (external)

3 Why is it necessary for an artery to go to the heart muscle itself?

4 Where do you think the arteries X, Y and Z are taking blood?

Blood returns to the heart via the superior and inferior vena cava.

5 Where has the blood come from that travels in these veins?

Contraction and relaxation of the heart muscle produce pressure changes within it. Look at Fig. 6.21. The graph shows the changes in pressure in the aorta and left ventricle. Look at the curve for the left ventricle; notice that there is a recurring pattern in the graph. Each time the pattern repeats, a heartbeat has taken place. This is equivalent to one heart cycle. One heart cycle is 0.8 s. Check this on the graph. There is a repeating pattern for the aorta as well. Check this on the graph.

You should also notice from Fig. 6.21 that

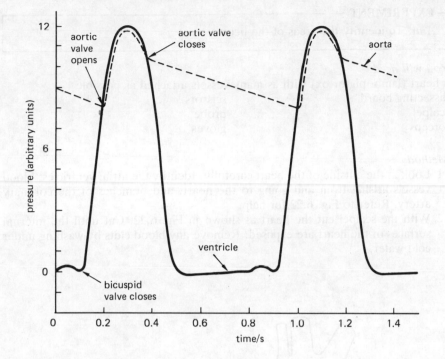

Figure 6.21 Graph showing pressure changes in the left ventricle and aorta of an adult

- pressure increases dramatically in the ventricle once the bicuspid valve is closed
- as the ventricle relaxes pressure drops
- the aortic valve closes as blood falls back due to lack of pressure from the left ventricle.

Exercise

Look at Fig. 6.21 again.

How long in one heart cycle is the pressure in the aorta greater than in the left ventricle?
What was the highest pressure recorded in the left ventricle?
What was the highest pressure recorded in the aorta?
What was the lowest pressure recorded in the left ventricle?
What was the lowest pressure recorded in the aorta?
Why does the pressure in the aorta remain high compared with the left ventricle?

⚠️

EXPERIMENT ─────────────────────────────────

Aim: To identify the areas of the heart

You will need:
a heart (lamb, pig or ox) with as many vessels attached as possible
dissecting board   scissors
scalpel   probe
forceps   gloves

Method
1 Look at the outside of the heart carefully. Identify the atria, ventricles, blood
  vessels arising from and going to the heart, and branches of the coronary
  artery. Refer to Fig. 6.22 for help.
2 With the scalpel cut the heart as shown in Fig. 6.22. Cut until the internal
  surfaces of the heart are exposed. Remove any blood clots by washing under
  cold water.

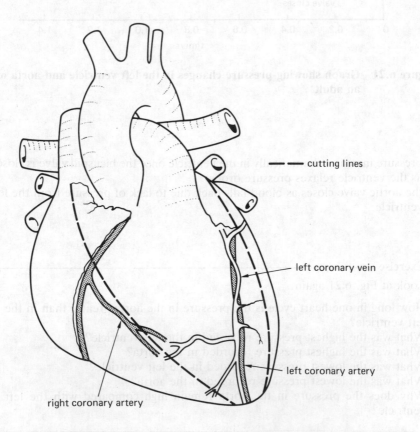

Figure 6.22   Dissection of the heart

3 Identify the atria and ventricles. Note the thicker wall of the left ventricle. (If your heart was not in the position shown in Fig. 6.22 the 'left side' may appear on the right.)

4 Find the tri- and bi-cuspid valves. Use your forceps and probe to examine them. If you cannot find three flaps on the right and two flaps on the left, you may have cut through one of them. ('Tricuspid' refers to the three cusps or flaps and the valve and 'bicuspid' refers to two cusps or flaps.)

5 Identify the chordae tendinae and notice the strong muscles supporting them.

6 Pull the chordae tendinae with the forceps. Feel how strong they are.

7 Feel the inside of the heart with your fingers and note how smooth it is.

8 Find the pulmonary artery and aorta (even in incomplete hearts the semilunar valves from these vessels often remain). Find the semilunar valves – they form three neat cusps. Note that these valves and the tri- and bi-cuspid valves all lie on the same plane through the heart.

9 Find the vena cava and pulmonary veins, if they are still attached to your heart. Can you find any differences in wall thickness and elasticity between the vena cava and the aorta?

10 Finally try to trace the route taken by blood as it travels through the heart.

Questions

1 Why is the wall of the left ventricle thicker than the wall of the right ventricle?

2 Why do you think the chordae tendinae are so strong?

3 Why is it important that the inside of the heart is smooth? If it was not smooth, what would happen to the blood as it passed through it?

EXPERIMENT

Hypothesis to be tested: Heart rate is affected by physical activity

---

You will need: a stopclock
a volunteer

Method

1 Count the pulse rate of your volunteer for 30 seconds. Double the number to get the pulse rate for one minute.

2 Ask your volunteer to exercise for two minutes (running on the spot, running around a building or using an exercise bike would all be suitable).

3 Immediately after the exercise count the pulse for 30 seconds. Double the number to get the pulse rate for one minute.

You will need to construct a results table for your results.

Questions

1 Was your volunteer's heart rate affected by physical activity?

2 Did you notice anything else different in your volunteer after exercise? Construct a hypothesis on the basis of your observations. Design an experiment to verify your hypothesis.

3 How long do you think your volunteer's heart was affected by physical activity? How could you modify this experiment to find out exactly how long it took for your volunteer to recover? How could you represent your results?

4 In the text it was stated that heart rate varies with age and fitness. Construct a hypothesis and design an experiment to see if this is so.
5 Can you think of any more experiments you could do in order to study an individual's response to exercise? Construct suitable hypotheses for these experiments.

Summary

■ The heart is a muscular pump.

■ The heart has four chambers: two *atria* and two *ventricles*.

■ *Valves* in the heart prevent backflow of blood.

■ The right side of the heart receives blood from the body. It pumps the blood to the lungs for oxygenation.

■ Oxygenated blood returns to the left side of the heart. It pumps the blood to the body via the aorta.

■ The coronary artery brings the heart muscle oxygen and nutrients.

■ The number of heartbeats per minute depends on the person's fitness and health.

Questions on Section 6.3

1 (a) How many valves are there in the heart?
  (b) What are the functions of these valves?
2 The pulses of an athlete and a typist are compared. The athlete's pulse is low compared to that of the typist.
  (a) Why is the athlete's pulse lower than the typist's?
  (b) The athlete and the typist run 100 metres. What effect would this activity have on their pulse rates?
3 (a) Why are the walls of both atria thin compared to those of the ventricles?
  (b) Why is the wall of the left ventricle much thicker than that of the right ventricle?

6.4 Serving the body

The circulation of the blood can be divided into

● the *pulmonary circulation*
● the *systemic* or *body circulation*.

We can say that the body has a double circulation. Look at Fig. 6.23. Notice that the systemic circulation includes all the organs of the body apart from the lungs.

Arteries supply the organs with oxygenated blood. Look at Fig. 6.24 which shows some of the organs of the body and the arteries that serve them.

Veins take blood away from the organs towards the heart. Look at Fig. 6.25, which shows some of the organs of the body and the veins that take blood away from them.

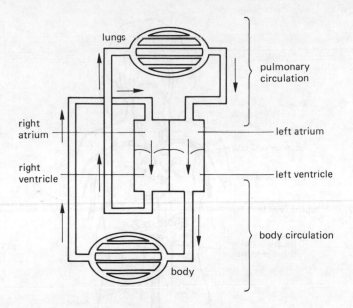

Figure 6.23   The double circulation

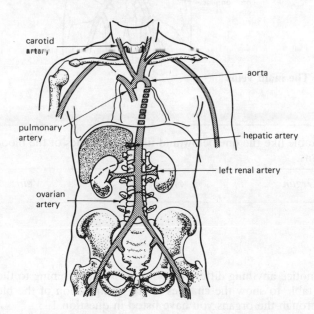

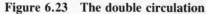

Figure 6.24   The main arteries

Compare Fig. 6.24 and Fig. 6.25. Notice that each organ has an *artery* taking blood to it and a *vein* taking blood away from it. As blood passes through each organ its composition changes. It may leave oxygen and collect carbon dioxide. It may leave glucose or urea, or it may collect these or other substances, depending on the organ.

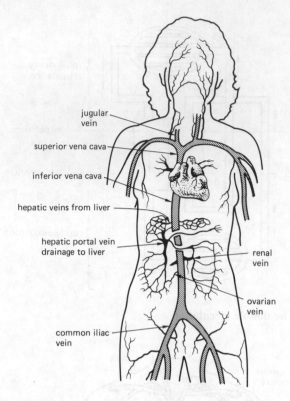

Figure 6.25   The main veins

Exercise

1 Make a table like the one below and fill in the names of the blood vessels and organs.

Organ	Artery	Vein

2 Do you notice anything different about the vessels coming to the liver?
3 Make a table to show the changes in the composition of the blood as it passes through the organs you have listed in question 1.

Summary

■ The body has a double circulation.

■ The circulation has two parts: the *pulmonary circulation* and the *systemic circulation*.

■ Every organ is associated with an artery and a vein.

■ The composition of blood changes as blood flows through the various organs.

Questions on Section 6.4

1 How does blood entering the liver via the hepatic artery get back to the heart?
2 Blood in the hepatic portal vein is coming from the intestine.
  (a) Compare the levels of glucose and amino acids in the blood of the hepatic portal vein and that in the hepatic vein.
  (b) Will the blood in the hepatic portal vein contain high levels of oxygen? Explain your answer.

6.5 Blood groups

The blood group system is based upon the presence or absence of antigens associated with the membranes of the red blood cells.

There are four blood groups: A, B, AB and O. The type of antigen associated with a red blood cell gives the blood its group name:

Group	Type of antigen
A	A
B	B
AB	AB
O	none

You probably know that someone who needs a blood transfusion must only be given blood of a certain group. That is:

Blood group of person needing blood	Compatible blood
A	O and A
B	O and B
AB	O, A, B and AB
O	O

Notice that

● a person with group AB can receive blood of any group.
● anyone can receive group O blood.

So why are some blood groups incompatible with others?

Red blood cells are found in liquid plasma. Plasma contains certain *antibodies* which can react with the antigens on red blood cells. Blood from each group contains the following antibodies:

Group	Antibody
A	B
B	A
AB	none
O	A and B

These antibodies work in the same way as one of the antibodies produced by the lymphocytes: that is, they cause red cells from incompatible blood to stick together. Let's see what happens when two incompatible bloods mix. This is shown in Fig. 6.26.

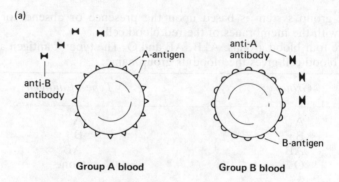

(a)

anti-B antibody

A-antigen

Group A blood

anti-A antibody

B-antigen

Group B blood

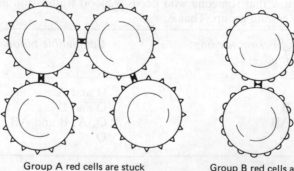

(b) The two bloods mix, i.e. a group A person has a transfusion of group B blood

Group A red cells are stuck together by anti-A antibodies from B blood

Group B red cells are stuck together by anti-B antibodies from A blood

Figure 6.26   The action between antigens on red blood cells and antibodies in plasma

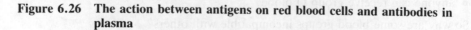

We can find which blood group a person belongs to by using samples of serum from blood groups A and B. (*Serum* is plasma from which the fibrinogen has been removed.) As these sera contain antibodies they are called *anti-sera* – anti-A and anti-B respectively.

One drop of blood of an unknown group is mixed with both anti-A serum and one with anti-B. Remember anti-B will clot *all* cells with antigen B on them. Anti-A will clot *all* cells with antigen A on them. Look at Table 6.1, which shows what happens when drops of known anti-sera are mixed with an unknown blood sample on microscope slides (a blood sample that is clotting has a speckled appearance).

Table 6.1 Determining unknown blood groups

Anti-sera used	Result on slides after adding unknown blood	Antibody in unknown blood	Blood group
anti-A	clotting	A	A
anti-B	no clotting	–	
anti-A	clotting	A	AB
anti-B	clotting	B	
anti-A	no clotting	–	B
anti-B	clotting	B	
anti-A	no clotting	–	O
anti-B	no clotting	–	

Exercise

Explain why blood group O is the 'universal donor'.
Explain why blood group AB is the 'universal receiver'.

Summary

■ There are four *blood groups*: A, B, AB and O.

■ The name of the blood group is determined by the type of antigen on the red blood cells.

■ Antibodies in blood plasma will stick foreign antigens together. So incompatible blood must not be used for transfusion.

■ This reaction between antigens and antibodies enables unknown blood groups to be determined.

—Questions on Section 6.5 ————————————————————

1 Why is it important to know the patient's blood group before giving a blood transfusion?
2 What would happen if group O and group AB blood came together?

6.6 **Rhesus factor**

Some people's red blood cells carry another antigen, which has nothing to do with blood groups. This is called the rhesus (Rh) factor. People who have the rhesus antigen on their cells are said to be rhesus-positive (Rh+). People who do not have the antigen are rhesus-negative (Rh−). This factor is only important if a rhesus-negative mother is carrying a rhesus-positive baby in her womb (uterus).

As you will find out in Section 11.4, the mother's blood and the baby's blood do not mix during the pregnancy. This is because their circulations are separated by an organ called the *placenta*. But when the baby is born the placenta comes away from the wall of the uterus, and at this time the baby's blood does come into contact with the mother's. Let's see what can happen. Look at Fig. 6.27.

Notice from Fig. 6.27 that the mother's body can only react to the different blood of the baby when the two bloods mix when the baby is born. Notice too that once her body has 'learnt' how to make Rh+ antibodies it will make them again whenever she comes into contact with the Rh+ antigen.

Now look at Fig. 6.28. This shows you that the solution to the rhesus problem is straightforward. The Rh+ antigen is made harmless by the injected antibodies, so the mother's body does not become sensitised at any stage.

—Summary ————————————————————————————————

■ Some people have an antigen associated with their red blood cells called the *rhesus factor*. They are rhesus-positive (Rh+).

■ Rh− blood does not have the antigen.

■ A Rh− mother can become sensitised to Rh+ antigens at the birth of an Rh+ child.

■ A sensitised Rh− mother will produce antibodies that will harm subsequent Rh+ babies.

■ An injection of antibodies at the birth of the Rh+ baby will stop the mother becoming sensitised and subsequent babies are unharmed.

—Question on Section 6.6 ————————————————————

During pregnancy, tests are carried out on the mother's blood. One of the tests is to see if she is rhesus-negative. Why is it important to find this out?

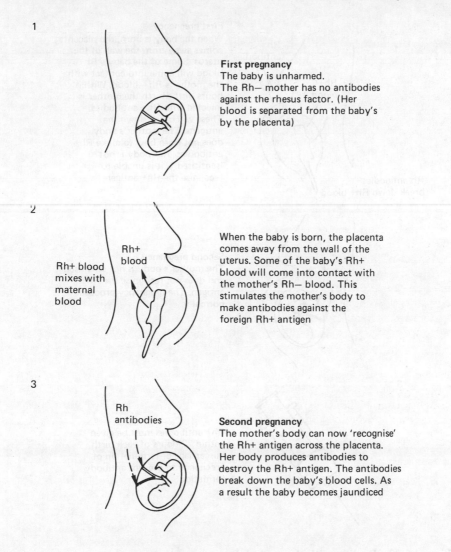

Figure 6.27   The possible result of being a rhesus-negative mother

6.7 What can go wrong?

Coronary heart disease, the commonest heart disease, is in fact a disease of the coronary arteries, which bring blood to the heart. In the UK, it is responsible for a third of the deaths in men over 40 years old. It kills fewer women, but the number of deaths among women is rising.

Fig. 6.29 explains how coronary heart disease develops. Notice that:

- the atheroma is caused by fatty deposits in the blood
- blood clotting in vessels may be a result of atheroma.

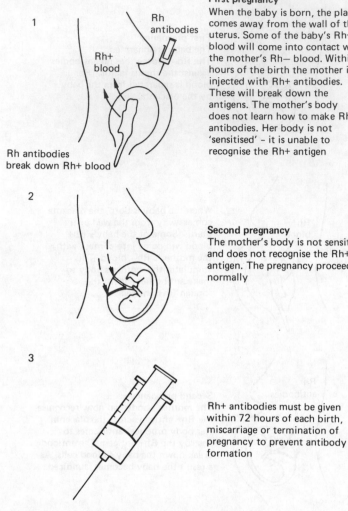

First pregnancy
When the baby is born, the placenta comes away from the wall of the uterus. Some of the baby's Rh+ blood will come into contact with the mother's Rh− blood. Within hours of the birth the mother is injected with Rh+ antibodies. These will break down the antigens. The mother's body does not learn how to make Rh+ antibodies. Her body is not 'sensitised' – it is unable to recognise the Rh+ antigen

1

Rh antibodies

Rh+ blood

Rh antibodies
break down Rh+ blood

2

Second pregnancy
The mother's body is not sensitised and does not recognise the Rh+ antigen. The pregnancy proceeds normally

3

Rh+ antibodies must be given within 72 hours of each birth, miscarriage or termination of pregnancy to prevent antibody formation

Figure 6.28   Dealing with the problem

There is no single cause of coronary heart disease. But certain activities increase the chance of getting the disease, and are known as *risk factors*. The risk factors are listed below.

Smoking cigarettes   Look at Fig. 6.30. Notice that

- the two substances in cigarette smoke affecting the health of the heart are *nicotine* and *carbon monoxide*
- the nicotine encourages the development of atheroma
- the ability of the blood to carry oxygen is reduced; if the blood vessels are narrowed, still less oxygen reaches the tissues.

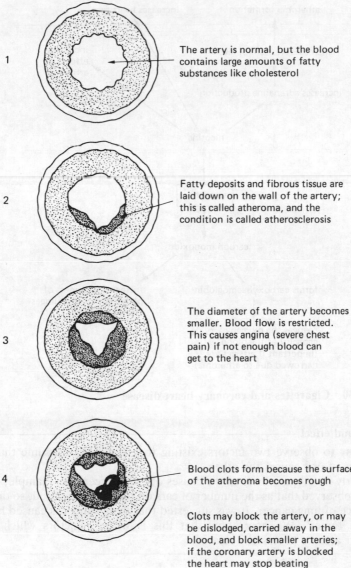

1 The artery is normal, but the blood contains large amounts of fatty substances like cholesterol

2 Fatty deposits and fibrous tissue are laid down on the wall of the artery; this is called atheroma, and the condition is called atherosclerosis

3 The diameter of the artery becomes smaller. Blood flow is restricted. This causes angina (severe chest pain) if not enough blood can get to the heart

4 Blood clots form because the surface of the atheroma becomes rough

Clots may block the artery, or may be dislodged, carried away in the blood, and block smaller arteries; if the coronary artery is blocked the heart may stop beating

Figure 6.29 Developing coronary heart disease

Cholesterol and lipids Cholesterol is used by the body for many purposes: to make cell surface membranes and male and female sex hormones, for example. It is also the main component of atheroma. High blood cholesterol levels are often associated with this atheroma, but this is not necessarily a 'cause and effect' relationship (see the boxed note 'Cause and effect').

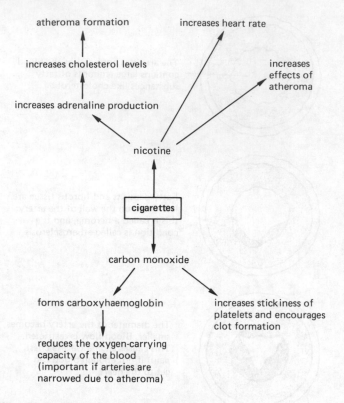

Figure 6.30   Cigarettes and coronary heart disease

Cause and effect

It is easy to observe two factors existing together and to assume that the presence of one causes the other.

In early investigations as to the causes of lung cancer, for example, some people observed that as the number of cars on the road increased, so did the incidence of lung cancer. It was suggested that 'lung cancer is caused by car exhaust fumes'. We now know that this is wrong. Factors which exist together are not necessarily related.

Cholesterol levels can be reduced by cutting down on animal fat, and foods that contain it (mainly eggs, butter, cream, suet, lard, and other meat fat), by eating more fruit and vegetables, and by replacing animal fats with vegetable oils (see the boxed note 'Saturated *vs* unsaturated').

Saturated *vs* unsaturated

Lipids may be

- *saturated* – solid, mostly animal
- *monounsaturated* – liquid, mostly vegetable
- *polyunsaturated* – liquid, mostly vegetable

Saturated lipids raise blood cholesterol levels and may therefore contribute to atheroma formation. Monounsaturated lipids do not affect blood cholesterol level. Polyunsaturated lipids will lower blood cholesterol levels.

Hydrogenated vegetable oils, which are used to make margarine, have been artificially saturated and behave like saturated lipids. The softer the margarine, the less hydrogenated (or saturated) it will be.

Stress increases levels of a hormone called adrenaline in the blood. Adrenaline releases fatty acids which can be broken down and supply the body with energy. If physical exercise does not follow the release of the fatty acids, the liver will convert them to cholesterol. The result is an increase in blood cholesterol levels.

Physical inactivity Exercise reduces the chance of developing coronary heart disease, especially exercise which does not make too great a demand on the body. For example, weight-lifting and press-ups increase blood pressure (see the boxed note 'High blood pressure and heart failure'). Jogging, swimming, cycling, walking and tennis are much better.

High blood pressure and heart failure

Atherosclerosis is associated with high blood pressure. High blood pressure causes a thickening in the left ventricle. Its demands for oxygen exceed its supply and the person becomes breathless. Failure of the left side and sometimes the right side of the heart may result.

Sex More men than women suffer from coronary heart disease. Women are protected by their hormones until after the menopause.

Heredity Coronary heart disease tends to run in families, but a carefully controlled diet can minimise the effect of heredity.

Diabetes Atherosclerosis is common in diabetes because the disease is often associated with obesity, high blood pressure and high blood cholesterol levels. Obese people are themselves more prone to high blood pressure and high cholesterol levels, but obesity on its own is not a risk factor.

Salt encourages the body to retain water. There is a relationship between a high salt intake and heart failure, but it is not a direct causal relationship.

Soft water More people die of heart disease in soft water areas than in hard. It is difficult to find out why this is so. We know that hard water does not dissolve toxic trace elements like cadmium and lead from plumbing, but we do not know if these elements affect the heart.

Alcohol In long-term persistent drinkers the heart may become enlarged and eventually fail. In addition, a large intake of alcohol may poison the heart directly.

Coffee It is difficult to find a direct link between coffee intake and heart disease, but drinking a lot of coffee can produce abnormal heart rhythms.

Oral contraceptives The risk to a woman who takes 'the pill' is slight unless she smokes, or has high blood pressure or a high level of blood cholesterol. There is a link between high blood pressure and oral contraceptives. The risk increases with the time over which the woman has taken the pill, and for women over 35 years of age.

Summary

- *Atheroma* is a fatty deposit in an artery.

- Atheroma leads to blockages and sometimes blood clots in the blood vessels.

- Atheroma is probably caused by several interacting factors. These are *risk factors*.

- The risk factors include cigarettes, cholesterol and lipids, stress, physical inactivity, sex, heredity, diabetes, salt, soft water, alcohol and oral contraceptives.

Questions on Section 6.7

1 A study showed that heart attacks were more common among drivers of double-decker buses than among conductors, who were walking up and down stairs all day. Why do you think this should be so?

Suggest a hypothesis which could be tested by experiment, based on this observation.

2 Since 1960 there has been an increase in heart attacks in younger women. Why do you think this is so?

3 Why are business executives more prone to coronary heart disease than people in some other jobs?

What advice would you give business executives wanting to cut their chances of developing the disease?
---

7 Breathing and respiration

The function of the breathing system is to move air in and out of the body and to enable gaseous exchange to take place. The term *gaseous exchange* describes the movement of oxygen from the lungs into the blood, and the movement of carbon dioxide from the blood into the lungs. Remember that *respiration* occurs in cells and that *breathing* refers to the action of the breathing system.

■ **Memory check**
What is respiration?
See Section 1.5.

7.1 The breathing system

Look at Fig. 7.1. Notice that

- air must be brought into the nasal cavity and into the *trachea*
- from the trachea air will pass into the *bronchus*
- the trachea and bronchus are kept open by C-shaped rings of cartilage
- from the bronchus air will pass into the *bronchioles*
- in the larger bronchioles the rings of cartilage are replaced by *plates of cartilage*, but there is no cartilage in bronchioles less than 1 mm across
- from the bronchioles air will pass into the *air sacs*
- gaseous exchange will take place in the air sacs. The many small air sacs provide a very large surface area for this purpose. The surface area of the air sacs in humans is estimated to be about 92 m^2.

Look at Fig. 7.2. This is a section of lung tissue as seen under a microscope. Note that the air sacs do not appear round; this is because they are compressed on all sides by other air sacs.

Let's look at the passages leading to the air sacs in a little more detail. The nasal passages and trachea clean, moisten and warm the air coming into the body. The passages have a supply of mucus-producing cells.

■ **Memory check**
Where have you come across mucus-producing cells before?
What was the function of the mucus they produced?
See Section 5.2.

Look at Fig. 7.3. Notice that dirt and microbes stick to the mucus. The carpet of *cilia* (microscopic hairs) move in a wave-like manner, taking the mucus and dirt towards the nose or to the back of the throat where it can be swallowed. Coughing also helps mucus and dirt to reach the back of the throat. These processes prevent dirt and microbes from entering the lungs.

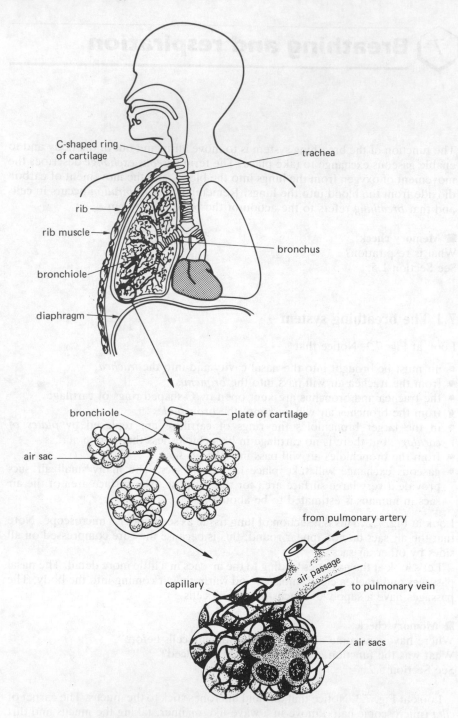

Figure 7.1   The breathing system

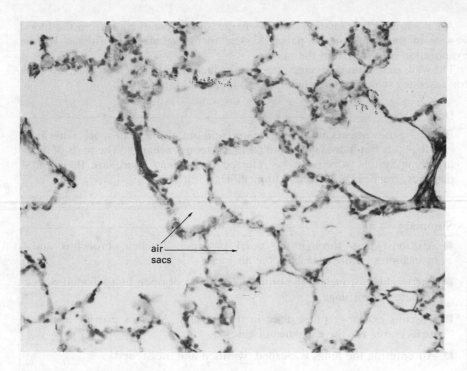

Figure 7.2   Section of lung tissue

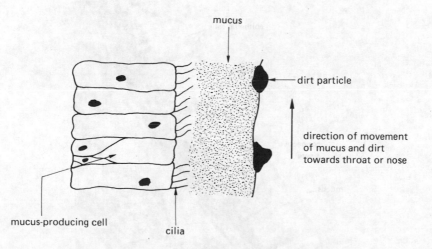

Figure 7.3   Detail of the lining of the trachea or bronchus

Air is warmed by radiation of heat energy and by conduction from the blood vessels in the walls of the passages. The incoming air is moistened by the evaporation of water from the mucus.

The walls of the nasal passages, trachea, bronchus and bronchioles are too thick for gaseous exchange. This region is known as *dead space*.

Dead space

Breathing movements must bring air into and out of the dead space. This air is in the body but it does not play a part in gaseous exchange. The parts of the breathing system which can be referred to as dead space are the nasal passages, trachea, bronchi and bronchioles.

Summary

■ Air must pass through the nasal passages, trachea, bronchus and bronchioles before it reaches the air sacs.

■ The trachea, bronchus and bronchioles are kept open by C-shaped rings or plates of cartilage.

■ Gaseous exchange takes place in the air sacs. Air sacs provide a large surface area for the exchange of gases.

■ Air entering the lungs is cleaned, warmed and moistened.

■ Mucus traps dirt and microbes coming into the air passages. Cilia move the mucus and dirt out of the passages.

Questions on Section 7.1

1 Look at Fig. 7.4.

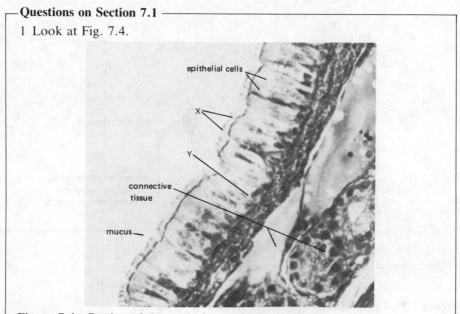

epithelial cells

X

Y

connective tissue

mucus

Figure 7.4   Section of the wall of a trachea

(a) What are the structures labelled X?
(b) What is the function of X?
(c) The cells labelled Y produce mucus. What is the function of the mucus?

2 Fig. 7.5 shows two photographs of lung tissue. One is a healthy lung and the other is unhealthy. Which is the healthy lung and which is the unhealthy lung? Give reasons for your choice.

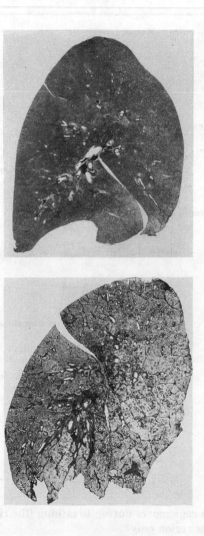

Figure 7.5

7.2 **Breathing in and breathing out**

Exercise

This is a short practical exercise.

1 Put the palms of your hands on either side of your rib cage.
2 Breathe in deeply.
3 Breathe out.
4 Breathe in and out again, taking note of what is happening to your ribs.

You should notice that your ribs are going *up* and *outwards* when you breathe in. Try the exercise again if you didn't notice these movements.

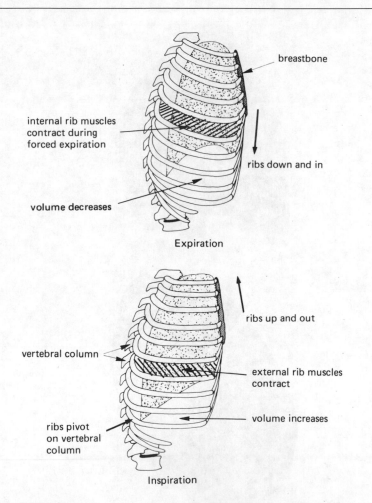

Figure 7.6   How the rib cage moves during breathing (the rib muscles are shown in one region only)

As your rib cage is moving up and outwards, the space inside your chest cavity is getting bigger. Air will be drawn into your lungs to fill the space.

Look at Fig. 7.1 again. Find the *ribs*. Notice the *rib muscles* between them. These muscles contract to lift the ribs up. When they relax the ribs go down.

The ribs can be pulled upwards because they form movable joints with the vertebrae. Look at Fig. 7.6. Find the ribs, vertebral column and breastbone. The action of the ribs can be compared with the handle of a bucket. Look at Fig. 7.7. Notice that moving the bucket handle up creates a space.

Look at Fig. 7.1 once more. Find the *diaphragm*. Notice that the diaphragm separates the chest cavity from the abdominal cavity. It is a sheet of strong connective tissue with muscle around its edges. When you breathe in the muscle contracts, pulling the diaphragm flat. This action makes the chest cavity larger, because it pushes down the abdominal contents. When you breathe out the diaphragm muscle relaxes. The contents of the abdomen push upwards, so that the diaphragm is pushed into a dome shape and air is pushed out of the lungs.

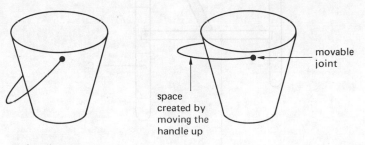

handle down

Figure 7.7 Bucket handle model of rib action during breathing

Describing breathing

Other words which can be used to describe breathing in and breathing out are:

Breathing in
inhalation
inspiration

Breathing out
exhalation
expiration

Summary

	■ *Breathing in*	■ *Breathing out*
ribs	go up and out	go down and in
muscles	contract	relax
diaphragm muscle	contracts	relaxes

| diaphragm | pulled flat so that it pushes down on the abdominal contents | becomes dome-shaped and pushes against the lungs |

Questions on Section 7.2

1 Look at Fig. 7.8. This is a model which can be used to show the movement of the ribs during breathing in and breathing out.

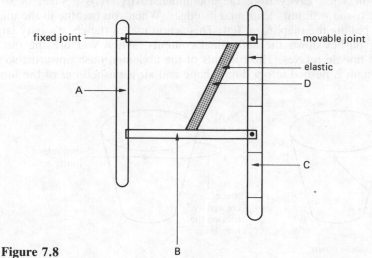

fixed joint

movable joint

elastic

A

D

C

Figure 7.8

B

(a) What do the structures labelled A, B, C and D represent?
(b) Explain how this model could be used to show the movement of the ribs during breathing in and breathing out.

2 Look at Fig. 7.9. This apparatus represents structures in the thorax associated with breathing, and can be used to demonstrate how movement of the diaphragm makes the lungs inflate.

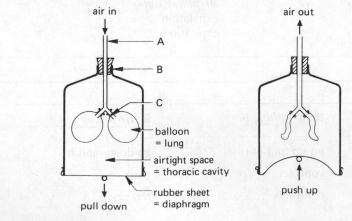

air in

air out

A

B

C

balloon
= lung

airtight space
= thoracic cavity

rubber sheet
= diaphragm

pull down

push up

Figure 7.9

(a) Explain how this apparatus can be used to demonstrate the action of the diaphragm in breathing.

(b) What structures are represented by the letters A, B and C?

(c) This apparatus is quite a poor representation of the structures in the thorax. Why?

7.3 Gaseous exchange

Look at Fig. 7.10, which is a simple diagram of an air sac. Find the wall of the air sac. Notice that there is a film of moisture around it. Notice that:

- the capillary surrounds the air sac
- the capillary wall is only one cell thick, so the red blood cells are as close as possible to the air in the air sacs
- oxygen diffuses across the wall of the air sac and combines with the haemoglobin of the red blood cells to make *oxyhaemoglobin*
- carbon dioxide diffuses out of the blood and moves across the wall of the air sac.

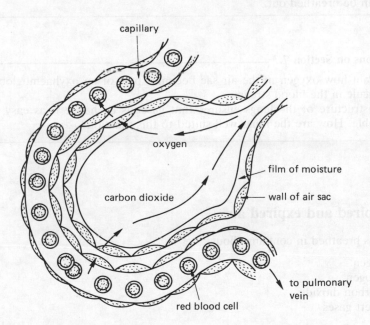

Figure 7.10 An air sac in the lung

■ **Memory check**

What is diffusion?

See Section 2.1.

Exercise

1 The wall of the air sac is very thin. What is the advantage of this?
2 Why should there be a good blood supply to the air sacs?
3 Why should there be a large surface area in the lungs?

Summary

■ Oxygen diffuses across the wall of the air sac and combines with haemoglobin in the red blood cells.

■ The air sacs provide a very large surface area for the exchange of gases.

■ The air sacs have thin walls so that gases can pass across them easily and quickly.

■ The air sacs have a good blood supply so that ample oxygenated blood is taken to the tissues, and also so that a diffusion gradient is maintained between the blood and the air in the air sacs.

■ Carbon dioxide diffuses across the wall of the air sac into the lung so that it can be breathed out.

Questions on Section 7.3

1 Explain how oxygen in the air sac becomes part of an oxyhaemoglobin molecule in the blood stream.
2 The structure of the lungs ensures that gaseous exchange is as easy as possible. How are the lungs well suited to this function?

7.4 Inspired and expired air

Air that is breathed in contains about

20% oxygen
79% nitrogen
0.04% carbon dioxide
0.16% inert gases

There are also variable amounts of water and pollutants, such as carbon monoxide, in air. Exhaled air contains about

16% oxygen
79% nitrogen
 4% carbon dioxide

as well as inert gases and water vapour.

Exercise

Cobalt chloride paper is used to test for the presence of water. It is blue when dry and pink when wet. Consider the following hypothesis: 'The air we breathe out contains more water vapour than the air we breathe in.'

Design an experiment to test this hypothesis, using cobalt chloride paper and any other apparatus you may need. Describe

- what apparatus you would need
- how you would carry out the experiment
- what results you would record
- any experimental errors which might lead to inaccurate results.

EXPERIMENT

Hypothesis to be tested: Expired air contains more carbon dioxide than inspired air does

---

You will need:
limewater
apparatus shown in Fig. 7.11

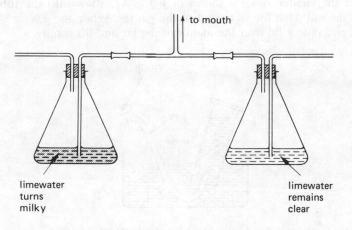

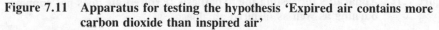

Figure 7.11   Apparatus for testing the hypothesis 'Expired air contains more carbon dioxide than inspired air'

Method
1 Pour limewater into both flasks so that they are half full.
2 Gently breathe in and out through the mouthpiece.
3 The air will enter the apparatus through one flask and leave by another. Try to work out which is which. Take note of what happens to the limewater.

Questions
1 Study the apparatus shown in Fig. 7.11.
2 Which flask of limewater has inspired air passing through it?
3 Which flask of limewater has expired air passing through it?
4 What has happened to the limewater in the flask through which expired air has passed?
5 What test for carbon dioxide have you learnt from this experiment?

┌─ **EXPERIMENT** ───

Hypothesis to be tested: Expired air does not support burning as well as inspired air does

---

You will need:
2 gas jars and greased lids   candle
trough of water   stopclock
rubber tube

Method
Read through the method, then construct a results table before you start the experiment.

1 Fill one gas jar with water and place it in the trough of water, so that the mouth of the jar is submerged.
2 Arrange the rubber rube as shown in Fig. 7.12. Blow into the tube. The expired air will push the water out of the gas jar. When the gas jar is full of expired air, slide a lid over the mouth of the jar and lift it out.

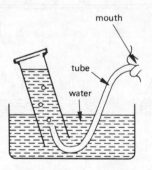

Figure 7.12   Apparatus for testing the hypothesis 'Expired air does not support burning as well as inspired air does'

3 Light a candle. Quickly remove the lid and place the gas jar of expired air over the candle. Start the stopclock immediately. Record how long it takes for the candle to go out.
4 Light the candle again. Repeat the experiment but this time use the other gas jar which will be full of ordinary air – that is, the air you breathe in, or inspired air. Record how long it takes for the candle to go out.

Questions
1 Did the expired air support burning as well as the inspired air did?
2 Can you think of any sources of error in this experiment?

EXPERIMENT

Hypothesis to be tested: Plants give out carbon dioxide in the dark

---

You will need:
5 test tubes
2 sprigs of *Elodea* (pond weed)
aluminium foil
lamp
bicarbonate indicator solution
1 drinking straw

Method
1 Set up the apparatus as shown in Fig. 7.13.
2 Leave the apparatus for 2 hours.
3 Record the colour of the bicarbonate indicator solution in each tube.
4 Use the straw and another tube of bicarbonate solution to find out what colour the solution will go with carbon dioxide.

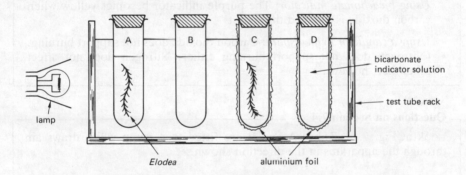

Figure 7.13   Apparatus for testing the hypothesis 'Plants give out carbon dioxide in the dark'

Questions
1 What is the test for carbon dioxide using bicarbonate indicator?
2 Do plants produce carbon dioxide in the dark?
3 What is the purpose of tube B?
4 Do plants produce carbon dioxide during the daytime? Explain the reason for your answer.
5 What is the purpose of tube D?

Summary

■ The differences between inspired air and expired air are summarised in the chart below:

Gas	Inspired air	Expired air
oxygen	more	less
carbon dioxide	less	more
water vapour	less	more

■ One test for water uses *cobalt chloride paper*. The paper is blue when dry and pink when wet.

■ Three tests for carbon dioxide are as follows:

Using limewater: The clear limewater becomes cloudy when carbon dioxide is bubbled through it.

Using bicarbonate indicator: The purple indicator becomes yellow when carbon dioxide is bubbled through it.

Using a candle or lighted splint: Carbon dioxide does not support burning. (Nitrogen does not support burning either. Nitrogen does not affect limewater, however.)

Questions on Section 7.4

Look at Fig. 7.14. Tubes A and B are fitted to a pump which draws air through the apparatus in the direction shown.

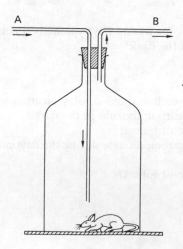

Figure 7.14

5 cm

1 How could you modify this apparatus to investigate whether the expired air from the mouse contained carbon dioxide?
2 What further modifications could you make to investigate whether the air entering the flask contained less carbon dioxide than the air leaving it?
3 Small droplets appear on the inside of the glass. It is assumed that these droplets are water. How could you test this?
4 Suggest a control for this experiment.

7.5 Cellular respiration

You learnt in Section 1.5 that respiration takes place in the cells of the body and results in the release of energy. The term *aerobic respiration* refers to respiration that uses oxygen. To respire aerobically a cell needs *oxygen* and *glucose*.

■ **Memory check**
How do oxygen and glucose get to the cells of the body?
See Section 6.4.

We are now going to study how aerobic respiration takes place in cells. Look at Fig. 7.15. Glucose and oxygen pass across the cell surface membrane and into the cytoplasm. A series of enzyme reactions takes place in the cell, resulting in the release of energy. The waste products from the reaction, water and carbon dioxide, pass out of the cell. These substances pass into the blood stream and will be eliminated from the body. (Read the boxed note 'Storing the energy'.)

― **Storing the energy** ―――――――――――――――――――――――――
The energy released in respiration is stored in molecules of ATP (adenosine triphosphate). ATP provides energy for the chemical processes going on in cells. It can release the stored energy very quickly.

We can summarise aerobic respiration thus:

glucose + oxygen $\longrightarrow$ carbon dioxide + water + energy

― **Exercise** ―――――――――――――――――――――――――――――――――
What do you think a cell needs energy for?

Look at Fig. 7.16(a). Notice that as energy demand increases, more oxygen is supplied to the cells. Now look at Fig. 7.16(b). Here the energy demand is increasing but the oxygen supply remains constant. This can easily occur in overworked muscles. The body copes with this for a short time by releasing energy from food without oxygen. This is *anaerobic respiration*. It can be summarised as:

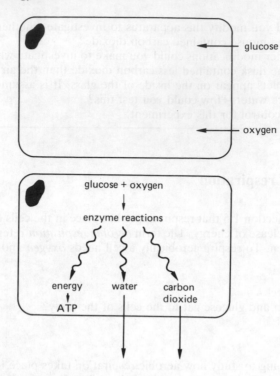

Figure 7.15 Aerobic respiration

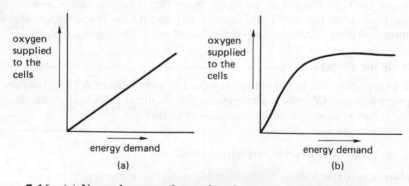

Figure 7.16 (a) Normal energy demand and oxygen supply; (b) high energy demand and low oxygen supply

glucose $\longrightarrow$ lactic acid + energy

The lactic acid passes into the blood stream and is taken to the liver. In the liver it is oxidised to carbon dioxide and water. This oxidation will continue after the exercise has stopped. But getting energy without oxygen is like getting energy on credit; it creates an 'oxygen debt'. Rapid and deep breathing after exercise has stopped supplies the oxygen for the oxidation of lactic acid. This is called *paying the oxygen debt*.

Exercise

In Section 6.3 you learnt that pulse rate increases with physical exercise.

1 Design an experiment to test the hypothesis 'Exercise causes breathing rate and depth to increase.'
2 Design an experiment to investigate how long it takes to repay an oxygen debt.

EXPERIMENT

Hypothesis to be tested: Yeast cells respire anaerobically

You will need:
2 boiling tubes
bung fitted with delivery tube
2 g dried yeast
5 g glucose
pasteur pipette

glass rod
oil
distilled water
water bath set at 30 °C
limewater

Method
1 Fit one boiling tube with the bung and delivery tube.
2 Put the yeast and glucose into the boiling tube.
3 Add 15 cm^3 distilled water to the boiling tube. Mix gently with the glass rod.
4 Pipette a little oil over the yeast mixture.
5 Set up the rest of the apparatus as shown in Fig. 7.17.

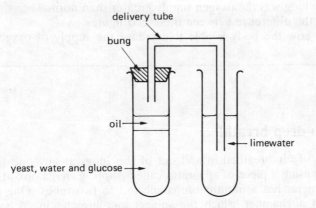

Figure 7.17 Apparatus for testing the hypothesis 'Yeast cells respire anaerobically'

6 Place the apparatus into a water bath at 30 °C. (The limewater does not have to be in the water bath. Water baths vary in size and design, so arrange the apparatus in any convenient way.)
7 Record any changes to the contents of the tubes.

Questions

1 Why was limewater used?
2 How does a change to the limewater indicate that the yeast has respired anaerobically?
3 What is the purpose of the oil?
4 Why was the tube containing the yeast placed in a water bath?
5 Suggest a suitable control for this experiment.
6 How could this experiment be modified so that the volume of carbon dioxide given off could be measured?

Summary

■ *Aerobic respiration*:

glucose + oxygen ⟶ carbon dioxide + water + energy

■ *Anaerobic respiration*:

glucose ⟶ lactic acid + energy

As anaerobic respiration is taking place an *oxygen debt* is being created. The oxygen debt is paid back by breathing deeply after the exercise has stopped.

Questions on Section 7.5

The graph in Fig. 7.18 shows oxygen supplied to the tissues before, during and after exercise.

1 For how long was the person exercising?
2 For how long was the oxygen supply greater than normal?
3 Explain the difference between these two figures.
4 Explain how the body is able to increase the supply of oxygen to the tissues.

7.6 Take a deep breath!

The amount of air breathed in and out of the lungs at any one time can be measured by using a piece of apparatus called a *spirometer*. Look at Fig. 7.19.

The spirometer has a mouthpiece connected to two tubes. One tube brings oxygen from a chamber which the subject can breathe in. A valve in the mouthpiece ensures that air breathed out goes along the second tube. This air is passed through soda lime (to absorb carbon dioxide) and then back to the chamber.

Breathing in and out causes the chamber to rise and fall; these movements are recorded by a pen drawing a trace on a piece of graph paper (kymograph).

The subject wears a noseclip, to ensure that all the air breathed in and out goes through the spirometer. It is possible to calibrate the graph paper so that the quantities of gases breathed in and out can be measured. The speed at which the

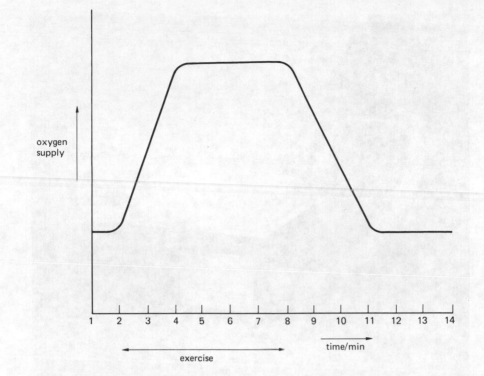

Figure 7.18 Oxygen supplied to the tissues before, during and after exercise

graph paper is moving can be regulated, and this allows the breathing rate and the rate of oxygen consumption to be calculated.

As the subject breathes in and out a trace is drawn on the graph paper. Look at Fig. 7.20(a). This trace was drawn while the subject was at rest: it is a measure of the *tidal volume*. Notice the regular pattern. As the graph paper is calibrated it is possible to measure how much gas was taken in with each breath. Look at Fig. 7.20(a) once more. Note that the subject breathed 6 times in 20 seconds. This gives a breathing rate of 18 times a minute.

If the subject was asked to jog for five minutes and then go on the spirometer, the trace would look like Fig. 7.20(b). Notice that both the *rate* and *depth* of breathing has increased.

Exercise

Look at Fig. 7.20(b).

1 What is the volume of air taken in at each breath?
2 What is the breathing rate of this person per minute?
3 You may notice that the trace is gradually sloping down to the right. Why do you think this is so?

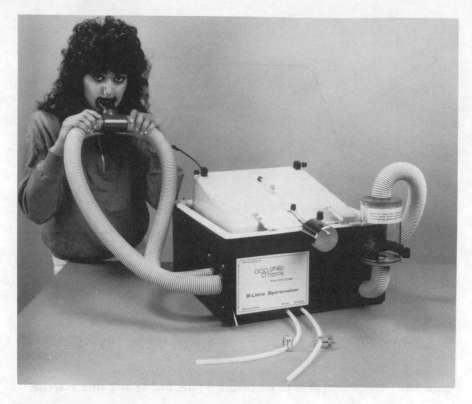

Figure 7.19   A spirometer in use

Look at Fig. 7.21. The person providing this trace was asked first to breathe in as deeply as possible, and then to breathe out as deeply as possible. The trace shows the extra volumes of air that were breathed in and out respectively. We call these two volumes *inspiratory reserve volume* and *expiratory reserve volume*.

Look at Fig. 7.21 again. The sum of inspiratory reserve, expiratory reserve and tidal volumes is called the *vital capacity*. This term describes the volume of lung space that is available for use if it is needed. But even when you have breathed out as much as you can, your lungs are not empty. A *residual volume* of air remains. Vital capacity and residual volume together form the *total lung volume*.

We can now calculate the volume of air breathed in per minute.

If a person's tidal volume is 500 cm^3
and the breathing rate is   16 times a minute, then

tidal volume × breathing rate =
500   ×   16   = 8000 cm^3 = 8 dm^3

We can also calculate the volume of oxygen breathed in per minute.

Air breathed in contains about 20% oxygen. If 8 dm^3 is breathed in per minute, the amount of oxygen breathed in is

$$20\% \text{ of } 8 \text{ dm}^3 = \frac{8000}{100} \times 20 = 1600 \text{ cm}^3$$

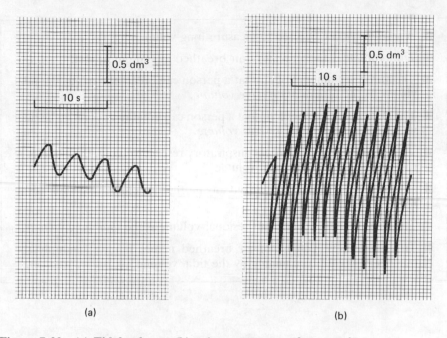

(a)   (b)

Figure 7.20   (a) Tidal volume; (b) spirometer trace after exercise

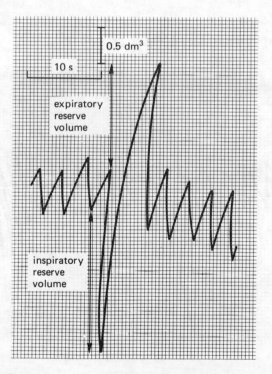

Figure 7.21   Inspiratory reserve volume, expiratory reserve volume and vital capacity

── Summary ────────────────────────────

- A *spirometer* can be used to measure lung volumes.

- *Tidal volume* is the amount of air breathed in and out at rest.

- The maximum amount of air that a person can breathe in minus the tidal volume is the *inspiratory reserve volume*.

- The maximum amount of air that a person can breathe out minus the tidal volume is the *expiratory reserve volume*.

- *Vital capacity* is the sum of the inspiratory reserve volume, the expiratory reserve volume and the tidal volume.

- When a person cannot breathe out any more, the volume remaining is the *residual volume*.

- The sum of vital capacity and residual volume is the *total lung volume*.

- To calculate the volume of air breathed in per minute, multiply the number of breaths per minute by the tidal volume.

── Questions on Section 7.6 ────────────

Look at the spirometer trace in Fig. 7.22.

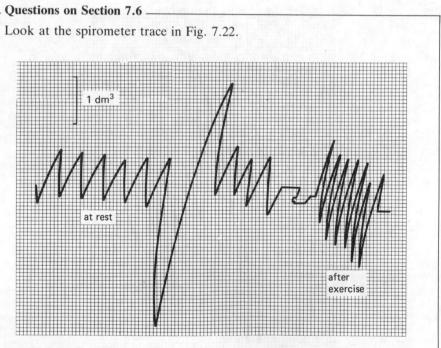

Figure 7.22

1 What is the tidal volume at rest?
2 What is the tidal volume after exercising?
3 What is the vital capacity?
4 The kymograph moved at a rate of 2.5 mm per second. How many breaths were taken per minute (a) at rest, (b) after exercise?
5 How many litres of air were inhaled per minute (a) at rest, (b) after exercise?

7.7 Artificial resuscitation

If there is insufficient oxygen in the blood, usually because breathing is abnormal or has stopped altogether, the tissues of the body become starved of oxygen. A condition known as *asphyxia* results.

Exercise

What might cause asphyxia?
Write down ten different possible causes as quickly as you can.

A person who is asphyxiated needs immediate revival or *resuscitation*. That is, normal breathing must be restored. If breathing has stopped you could try to resusciate the person. The two best-known methods are

- mouth-to-mouth resuscitation
- the Holgar–Nielsen method.

Mouth-to-mouth resuscitation

1 Open the airway by tilting the head backwards and pushing the chin upwards. (You may have to remove weed or other items from the mouth. Do not take out false teeth.) Look at Fig. 7.23(a). Notice that the tongue cannot block the airway if the head is tilted in this position.
2 Pinch the person's nose with your fingers. Take a deep breath and put your lips around the mouth. Blow into the lungs until the chest rises. Remove your mouth and allow the chest to fall. Count to five, and repeat. Go on until the person starts to breathe. Look at Fig. 7.23(b). Remember to keep the head tilting backwards all the time.

Mouth-to-nose resuscitation may be preferred if there is injury to the mouth.
The mouth-to-mouth method cannot be used if there is severe injury to the face and mouth, if the person is trapped facing downwards or if vomiting is interfering with breathing. In this case the Holgar–Nielsen method is used.

Holgar–Nielsen resuscitation

1 Look at Fig. 7.24(a). This is the correct position for the asphyxiated person and the resuscitator. Notice that the person lies face downwards, the arms

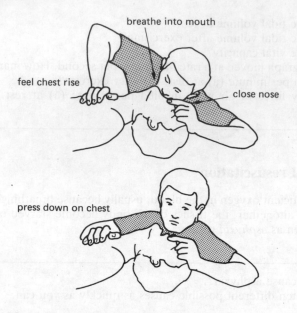

Figure 7.23   Mouth-to-mouth resuscitation.

over the head, and the elbows flexed so that one hand rests on another, with one cheek resting on the uppermost hand. Notice also where the resuscitator's knee and foot are placed.

2 Look at Fig. 7.24(b). The resuscitator's hands are placed on the back, below the shoulder blades. The resuscitator rocks forward with arms held straight, pushing downwards on the chest.

3 Look at Fig. 7.24(c). The resuscitator relaxes the pressure on the asphyxiated person's back, takes hold of the person's arms just above the elbows and raises them gently, and then returns them to the ground.

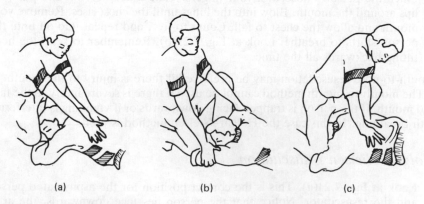

(a)   (b)   (c)

Figure 7.24   Holgar–Nielsen method of resuscitation

Note that:

- pressing down on the chest pushes air out of the lungs
- lifting the arms expands the chest cavity, so that air goes into the lungs.

The process of applying pressure to the lungs and then allowing them to expand should be repeated about 12 times per minute until the person starts to breathe.

How does resuscitation work?

In both methods of resuscitation air is forced into the lungs. Even expired air contains 16% oxygen, and this is enough to oxygenate the blood and thereby supply the tissues with oxygen.

Summary

■ *Mouth-to-mouth resuscitation* involves opening the airway, and inflating the asphyxiated person's lungs with expired air. The process is repeated about 12–15 times per minute.

■ *Holgar–Nielsen resuscitation* involves compressing the lungs by pressing on the chest cavity, and expanding the lungs by lifting the arms. The process is repeated 12 times a minute.

Questions on Section 7.7

1 During mouth-to-mouth resuscitation, the resuscitator is breathing out into the unconscious person. How does this:
 (a) maintain life in the tissues?
 (b) stimulate breathing movements?
2 Explain how the Holgar–Nielsen method gets air into the chest of the patient.

7.8 Smoking

You know that smoking seriously affects people's health. Smoking is banned altogether in many offices, restaurants and so forth. What is so harmful about smoking?

Let's start with tobacco smoke. It is a mixture of nearly a thousand substances. Some of them are gases; others form tarry droplets. They have various effects on the body:

- some are *irritants* and cause coughing
- some are *carcinogens*, which means they make cancers start to grow
- some are *cancer promoters* and speed up the growth of an established cancer
- some cause a breakdown in the defence mechanisms of the respiratory system which may lead to *chronic bronchitis* (read the boxed note 'Nicotine and carbon monoxide').

┌───┐
Nicotine and carbon monoxide

Nicotine affects the nervous system:

- it causes vasoconstriction (described in Section 8.1)
- it causes a rise in blood pressure
- it acts as a stimulant in small doses
- it acts as a depressant in large doses.

Carbon monoxide combines irreversibly with haemoglobin to form *carboxyhaemoglobin*. This reduces the ability of the blood to carry oxygen and the tissues become starved of oxygen. Carbon monoxide also increases the permeability of arterial walls to cholesterol and contributes to atherosclerosis (see Section 6.7).
└───┘

How smoking affects the breathing system

Inhaled smoke slows down the action of cilia in the trachea and bronchus, and may even stop their movement altogether. Eventually the ciliated cells may die. Look at Fig. 7.25(a). This shows cilia in a healthy bronchus. Fig. 7.25(b) shows the cilia in a bronchus exposed to cigarette smoke.

Secondly, the irritants in smoke make the mucus-producing cells in the lung produce more mucus. The mucus collects in the airways. The body tries to get it out by coughing. The 'smoker's cough' is the first sign of bronchitis.

At the same time, bacteria and other microbes get into the lower breathing passages like the bronchus. White blood cells try to engulf the bacteria. The activity may be so great that the surface layer of the bronchus may become damaged. Mucus is still being produced, and phlegm or pus – a mixture of white blood cells, cell debris, mucus and microbes like bacteria – collects; this makes the coughing worse.

Damage and inflammation of the bronchus contributes to *bronchitis*. Bronchitis is a serious illness which is responsible for over 30 000 deaths per year.

With the cilia out of action, mucus carrying dirt, microbes and tar slip further down into the air sacs. The only way the body can deal with these substances is for white blood cells to move through the wall and engulf them. Fibrous tissue may grow over them and seal them into the walls of the air sacs. They will stay there for life. Many of the substances in cigarette smoke cause cancer, so their accumulation increases the risk of cancer developing (see below).

Other effects of cigarette smoking

Emphysema
Bronchitis often leads to the breakdown of lung tissue, a condition called emphysema. The walls of the remaining air sacs become thickened. The result is a smaller surface area for gaseous exchange; moreover, gases cannot pass across the wall of the air sac so easily. So the lungs do not work as well as they should.

Cancer
Continued exposure to cigarette smoke stimulates the cells of the bronchi to divide and produce many layers of cells – in fact, they become more like skin cells

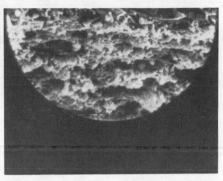

Figure 7.25 (a) Cilia in a healthy bronchus; (b) in a bronchus exposed to cigarette smoke, the cilia are no longer visible

in the way they behave. The increased rate of cell division increases the chances of cancerous cells being produced. Cancerous cells divide rapidly. The cells push their way through surrounding tissues and destroy existing structures. Some cells may break away and be carried in the lymph to other parts of the body, where they may form secondary cancers.

Coronary heart disease
You may remember from Section 6.7 that there are many factors related to this disease, and among them is the number of cigarettes smoked. In a heavy smoker, the inner lining of the coronary artery becomes thickened. Blood flow to the heart muscle may be reduced by as much as a half.

── EXPERIMENT ──────────────────────────────
Aim: To investigate what is contained in cigarette smoke

You will need:
the apparatus shown in Fig. 7.26

Method
1 Turn on the filter pump.
2 Light the cigarette.

Questions
1 What does the filter pump do?
2 What has collected on the glass wool?

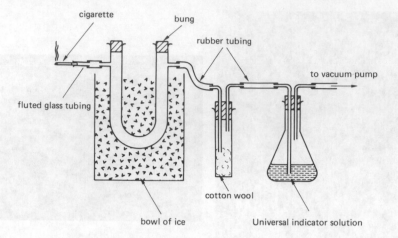

Figure 7.26   Apparatus for experiment to investigate what is contained in cigarette smoke

3 How could you tell if any substances have dissolved in the water? Give one simple test.
4 Can you suggest a suitable control for this experiment?
5 How could you compare the tar content of different cigarettes?
6 How could you compare the tar content of the first half of a cigarette with that of a whole cigarette?
7 How could you compare the tar content of tipped and non-tipped cigarettes?

── **Summary** ───

■ Cigarette smoke contains a mixture of substances which act as irritants, carcinogens and cancer accelerators, cause bronchitis and emphysema, and contribute to coronary heart disease.

── **PROJECT WORK** ────────────────────────────────

1 What is passive smoking? Find out how passive smoking can affect the non-smoker.
2 Why is smoking during pregnancy unwise? What advice would you give to a smoker who is pregnant?

7.9 Industrial diseases of the lungs

Air is not always clean. The cilia can usually clean the air coming into the body effectively, but if the amount of dust, chemicals and so on in the air is too great for the cilia to deal with, these substances can get into the air sacs. Ciliary action cannot remove soluble substances from the air, however. Soluble substances dissolve in the film of moisture lining the air sacs. The chemical can then move across the thin wall of the air sac and enter the blood stream. If it is a toxic substance, poisoning follows.

Insoluble substances in the air sacs are dealt with in three ways:

1 White blood cells move through the lungs, engulfing solid inhaled particles. They can then move to the cilia and are eventually swallowed.
2 If they cannot be moved out of the breathing system, the fibrous tissue grows over them and seals them into the wall of the air sac.
3 The fate of these particles is as follows:

   - they may stay in the air sacs
   - they may move into the lymphatic system and collect in the lymph glands
   - they may get into the blood stream.

Clearly, whether or not these particles cause disease depends on their chemical and physical nature. Let's look at two of the diseases that can be caused by particles that people may inhale at their place of work.

Asbestos

Asbestos is a fibrous mineral silicate. It is inert. That is, it does not easily react with anything; so it has been used extensively for building, insulation and fire protection. As it is fibrous it is easy to mould and manipulate.

The first cases of disease known to be caused by exposure to asbestos occurred in the 1900s. Since then, its production and use have been strictly regulated.

There are four types of asbestos, which differ in their chemistry. All are fibrous, and all contain fibres that can be inhaled. It appears that the fibrous nature of the substance is important in causing disease.

The disease caused by exposure to asbestos is called *asbestosis*. Asbestos fibres may penetrate the walls of the air sacs. The body makes fibrous tissue to repair the damage. White blood cells (phagocytes) try to engulf the asbestos fibres, but often these are bigger than the cells. The phagocytes are then damaged, and may leak their contents. Leakage from phagocytes appears to initiate the formation of more fibrous tissue in the lung (fibrosis). The air sac walls become very tough, thick and fibrous. A condition known as pneumoconiosis develops. Cancer may also arise in the area of the fibrosis. Lung cancer can occur with or independently of asbestosis. A smoker working with asbestos is 90 times more likely to get cancer than a non-smoker in the same job.

Coal

Exposure to coal dust may result in coal worker's pneumoconiosis. There are several forms of this:

- non-fibrous pneumoconiosis: there is no permanent damage and the wall of the air sac is intact
- progressive massive fibrosis, in which the tissues are permanently altered as a result of coal dust
- emphysema
- rheumatoid pneumoconiosis, in which large nodules of fibrous tissue develop. This is life-shortening.

Summary

■ Inhaling insoluble aerosols like asbestos and coal dust lead to fibrous tissue being laid down in the air sacs.

■ The ability to take up oxygen is reduced.

■ Emphysema and cancer may develop.

Questions on Section 7.9

Imagine you are a manager in a factory which handles a substance that causes irritation to the breathing system. Some evidence has come to light that it may be related to respiratory diseases.

1 How would you make the handling of the substance safer?
2 Would your method of safe handling be affected by whether the substance present in the air was soluble or insoluble?

8 Thermostats and waste disposal

In Section 1.5 you learnt that chemical reactions in cells produce *excretory products*.

■ **Memory check**
Can you list the excretory products?
What processes in the body produce them?
How are the products eliminated?
See Section 1.5.

The organs of excretion are the *lungs* (considered in the last chapter), the *skin* and the *kidneys*. In this chapter we are going to consider the skin and the kidneys.

8.1 The skin

Your skin is the largest organ in your body. It stops you drying out, repels invading microbes, protects you from physical damage, tells you about your environment, helps to regulate your body temperature and acts as an excretory organ.

The skin has a complex structure because it has so many functions. Look at Fig. 8.1. Notice that the skin is made up of two layers, the *epidermis* and the *dermis*.

The epidermis is the dead outer surface of your skin. You are losing the dead cells of the epidermis all the time. Clearly, if the outer surface is being worn away, new cells are needed to replace those lost. Find the *germinative layer* in Fig. 8.1. In this layer:

- cells are constantly dividing
- older cells are pushed towards the surface of the skin
- *melanocytes* produce a pigment *melanin*. Melanin ranges in colour from yellow to black. It protects the skin from the damaging effects of ultra-violet radiation in the Sun's rays. Sunlight stimulates the production of melanin.

Find the layer containing the melanocytes in Fig. 8.1.

Notice from Fig. 8.1 that the germinative layer surrounds the hair follicles and sweat glands and gives rise to the *granular layer*. Several things are happening to the cells in this layer:

- their nuclei are breaking down, which in time leads to the death of the cells
- the tough, strong protein *keratin* is formed in the cells of this layer
- the cells become flatter
- melanin is destroyed.

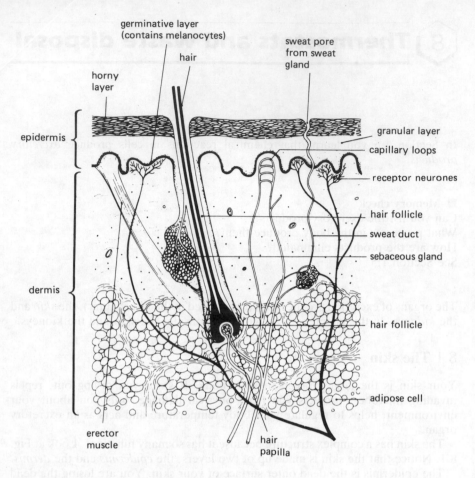

Figure 8.1　Section of the skin

The granular layer gives rise to the *horny layer*. Find this layer in Fig. 8.1. In this layer:

- the cells are dead
- they contain a high proportion of keratin
- they are gradually shed.

This keratinised layer makes the skin strong and waterproof, thereby protecting the body against water loss, injury and infection.

The efficiency of the horny layer is helped by an oily secretion from the *sebaceous glands*. Look at Fig. 8.1 and find the sebaceous gland shown there. The secretion is called *sebum*. Sebum

- makes the skin waterproof
- keeps the epidermis supple
- prevents dryness of the skin
- prevents growth of fungi and bacteria on the surface of the skin.

The dermis is the jelly-like layer lying below the epidermis. It has strong inelastic and elastic fibres running through it. The elastic fibres give the skin flexibility.

Look at Fig. 8.1 and find the layer of *adipose* cells (fat cells). The functions of this layer (adipose tissue) are

- to support the dermis
- to act as a food store
- to cushion the body against knocks and blows
- to act as an insulating layer.

Notice from Fig. 8.1 that the dermis is supplied with oxygen and nutrients from the blood vessels.

The dermis has two important functions: *temperature control* and *sensitivity*.

Temperature control

Body temperature is increased by:

- illness (fever)
- exercise (muscle contraction produces heat)
- wearing too many clothes
- high air temperature.

Look at Fig. 8.2(a) and (b). The blood vessels close to the surface of the epidermis can dilate or contract. If the vessels dilate, blood travels close to the surface of the skin and heat energy is lost by radiation. If the vessels contract, blood travels lower down in the skin and heat energy is retained.

Find the sweat glands in Fig. 8.1. When the blood is 0.5–1.0 °C warmer than normal, water, salts and a little urea pass from the blood and into the sweat glands. This mixture passes up the duct and on to the surface of the skin. The evaporation of sweat has a cooling effect on the skin.

What makes the blood vessels dilate or contract? and what makes the sweat glands start to secrete? Their actions in maintaining body temperature are controlled by a part of the brain called the *hypothalamus*. Look at Fig. 8.3 and

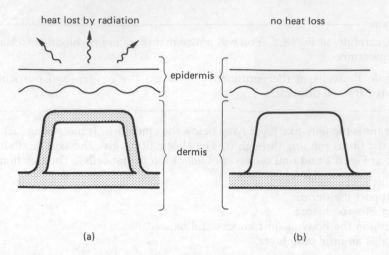

heat lost by radiation   no heat loss

epidermis

dermis

(a)   (b)

Figure 8.2   (a) Vasodilation (dilation of blood vessels near the surface of the skin); (b) vasoconstriction (narrowing of the same blood vessels)

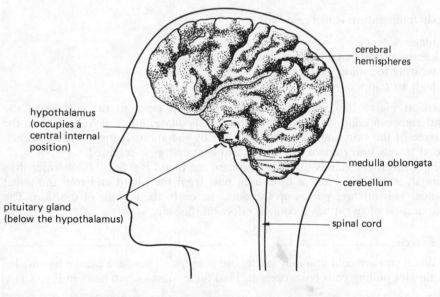

cerebral
hemispheres

hypothalamus
(occupies a
central internal
position)

medulla oblongata

cerebellum

pituitary gland
(below the hypothalamus)

spinal cord

Figure 8.3   The brain

find the hypothalamus. The hypothalamus is sensitive to the temperature of the blood that passes through it. It stimulates the temperature-regulating mechanisms like sweating and vasodilation if the blood temperature is high.

Sometimes the body is unable to maintain a constant body temperature and *hypothermia* results (read the boxed note 'Hypothermia').

— **Hypothermia** —————————————————————————————

Study the chart below.

Body temperature/°C	Effect on the body
37	normal body temperature
35	behaviour, speech and sight affected
32	unconsciousness which will result in death if the body is not warmed

When the body temperature falls to 35 °C or lower, the person is said to be suffering from hypothermia. The causes of hypothermia are:

- cold surroundings
- inactivity in cold surroundings
- insufficient food in cold surroundings
- experiencing cold and wet.

A person with hypothermia should be covered with a warm blanket and given sips of a warm drink.

Sensitivity

Look at Fig. 8.1. Find the receptor neurones. Some endings are high up in the dermis, like the one labelled in the drawing, and others are low down. Neurones whose endings are low down in the dermis are sensitive to pressure. Neurones whose endings are high up in the dermis are sensitive to touch, temperature and pain.

Notice the nerve net around the hair follicle. When an ant crawls up your leg, you usually feel it because it is moving the hairs and stimulating the nerve net around the follicle. So the hairs on your legs may not keep you warm but they help to keep you informed!

■ **Memory check**
Which vitamin is made in the skin when it is exposed to sunlight?
See Section 3.1.

EXPERIMENT

Hypothesis to be tested: Clothes keep us warm

You will need:
retort stand
boiling tube
thermometer
piece of cloth
bunsen burner

boss and clamp
measuring cylinder
stopclock
elastic band
safety spectacles

Method
Read the method through, and then construct a results table *before* you begin the experiment.

1 Measure 20 cm³ water and pour it into the boiling tube.
2 Support the tube on the retort stand.
3 Put the thermometer into the water.
4 Wearing safety spectacles, warm the boiling tube of water, using the bunsen burner. Stir the water gently with the thermometer. Continue heating until the water is about 40 °C.
5 Record the exact temperature.
6 Start the stopclock.
7 Record the temperature every 30 seconds for five minutes.
8 Heat the water to about 40 °C again.
9 Quickly put the cloth around the boiling tube. Support it with an elastic band, as shown in Fig. 8.4.
10 Record the exact temperature.
11 Start the stopclock.
12 Record the temperature every 30 seconds for five minutes.

Present your results from both tubes on a graph.

Questions
1 How does this experiment test the hypothesis that clothes keep us warm?
2 List any inaccuracies or errors in this experiment. Can the experiment be modified to eliminate these errors?
3 Design an experiment to find out what sort of fabric keeps us warm the best. State the hypothesis, the equipment you would need and how you would carry out the experiment. Construct a results table you would use.
4 It is said that many thin layers keep you warmer than one thick layer. Design an experiment to test this statement. State the hypothesis, the equipment you would need and how you would carry out the experiment. Construct a results table you would use.

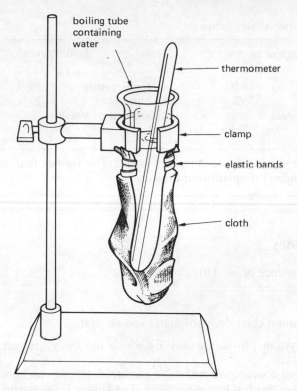

boiling tube
containing
water

thermometer

clamp

elastic bands

cloth

Figure 8.4   Apparatus for testing the hypothesis 'Clothes keep us warm'

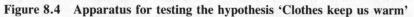

── Summary ────────────────────────────────

■ The skin is made up of two layers: the *dermis* and the *epidermis*.

Layer	Function
epidermis	melanin protects against u.v. light; keratinised, tough horny layer protects against injury and is waterproof
dermis	temperature control through vasodilation or vaso-constriction, sweating and shivering; information reception via sensitive nerve endings; sebum helps to prevent microbial growth and keeps the skin supple; elastic fibres give the skin elasticity

── Questions on Section 8.1 ─────────────────

1 After a game of football, the surface of a player's skin becomes red and moist.
   (a) What changes have occurred in the skin to bring about these effects?
   (b) What is the purpose of these changes?

2 Study the two tables below:

Inactive person		*Active person*	
temperature	18 °C	temperature	18 °C
humidity	35%	humidity	75%
amount of sweat lost	2 dm³	amount of sweat lost	. . . .

Estimate the missing figure in the right-hand table (notice that the humidity was much higher). Explain your answer.

8.2 The kidney

The kidneys produce *urine*. Urine contains

- water
- urea
- salts like sodium chloride, phosphates and sulphates.

The urinary system provides a way by which the body can get rid of these excretory products.

The volume of concentration of urine produced depends on how much fluid is taken in and how much is lost in sweating. The kidney is the means by which the body can control and balance the amount of water it contains.

You have two kidneys. They lie at the back of the abdomen, one on either side of the vertebral column just below the ribs. Look at Fig. 8.5. Take note of the average size of a human kidney. The *renal arteries* supply the kidneys with blood. These arteries arise from the aorta. *Renal veins* pass from the kidneys and drain

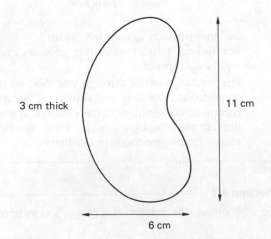

3 cm thick   11 cm   6 cm

Figure 8.5   The size and shape of a human kidney

into the inferior vena cava. Look at Fig. 8.6. Find the renal arteries and the aorta.
Find the renal veins and the inferior vena cava.

Look at Fig. 8.6 again. Find the *ureters*, *bladder* and *urethra*. Urine is formed in
the kidneys and passes through the ureters to the bladder. The bladder stores the
urine. Eventually urine flows out of the body via the urethra.

Look at Fig. 8.7. This shows a longitudinal section through a kidney. Notice the
three main regions of the kidney: the *cortex*, the *medulla* and the *pelvis*. Urine
drains into the pelvis and then passes into the ureters.

To understand how the kidney works we must consider its microscopic
structure. The kidney is made up of many similar tiny structures called *nephrons*.

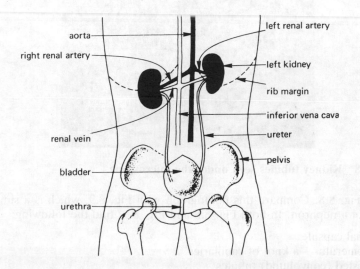

Figure 8.6   The urinary system (front part of pelvis not drawn)

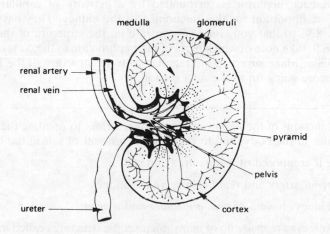

Figure 8.7   Longitudinal section of a kidney

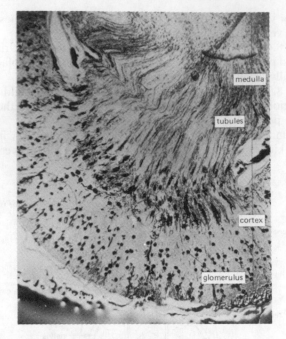

Figure 8.8   Kidney tubules seen under the microscope

Look at Fig. 8.8. Compare this photograph with Fig. 8.9, which is a simplified drawing of a nephron. In both Fig. 8.8 and Fig. 8.9, find the following:

- the renal capsule
- the glomerulus – a knot of capillaries
- the twisted (convoluted) tubules
- the loop of Henle
- the collecting duct.

Notice that each nephron is surrounded by a network of capillaries. These capillaries are important for the functioning of the kidney. They have been left out of Fig. 8.9, so that you could concentrate on the structure of the nephron. From Fig. 8.9, take note of which parts of the nephron are in the cortex and which in the medulla. Make sure you are familiar with the structure of the kidney and nephron before going on to the next section.

Summary

- The functions of the *kidney* are to produce urine, to regulate the amount of water in the body, and to regulate the amount of salt in the body.

- *Urine* is composed of water, urea and salts.

- The *renal artery* and *renal vein* serve the kidneys.

- The kidneys have a *cortex*, *a medulla* and a *pelvis*.

- The kdineys are made up of many microscopic structures called *nephrons*.

- Urine passes from the kidneys to the bladder via the *ureters*.

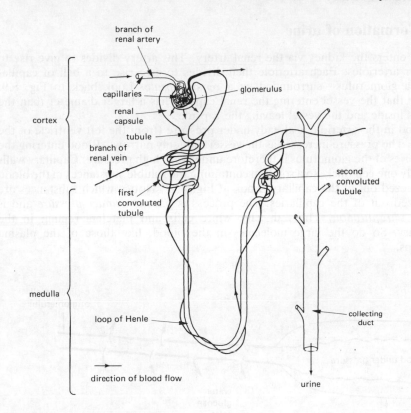

Figure 8.9   A kidney nephron

Look at Fig. 8.10.

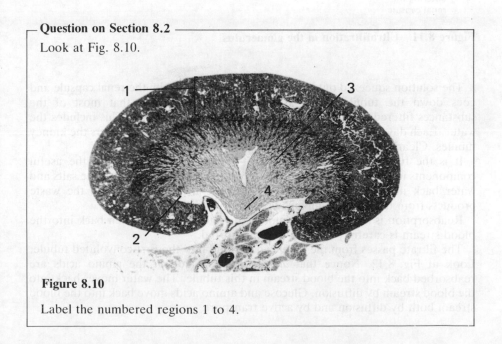

Figure 8.10

Label the numbered regions 1 to 4.

8.3 **Formation of urine**

Blood enters the kidney via the renal artery. This artery divides to give rise to smaller arterioles. Each arteriole then divides to give rise to a ball of capillaries – a glomerulus – surrounded by a *renal capsule*. Look back to Fig. 8.9. Notice that the vessel entering the renal capsule has a larger diameter than the vessels inside and the vessel leaving the capsule.

Blood in the arteriole is already under pressure (from the left ventricle of the heart). The pressure increases as the vessel suddenly narrows. Blood entering the capillaries of the glomerulus is therefore under very high pressure. Capillary walls are only one cell thick, so a solution containing the soluble substances in the blood is squeezed out of the capillary. Look at Fig. 8.11. Notice which substances are squeezed out of the capillaries. The process is *filtration under pressure* and is called *ultrafiltration*. The red cells, white cells and platelets remain in the capillary. So do the large molecules in the blood, like those of the plasma proteins.

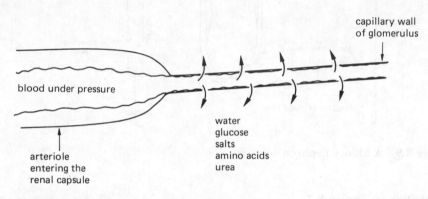

Figure 8.11   Ultrafiltration in the glomerulus

The solution squeezed out of the capillary is collected by the renal capsule and goes down the tubule. Look at Fig. 8.11 again. Notice that most of the substances filtered from the blood are still useful to the body. This includes the water. Each day 170 dm^3 of water is filtered from the blood and enters the kidney tubules. Clearly, the body cannot lose all this water in urine.

It is the function of the remainder of the nephron to *reabsorb* the useful components in the filtrate: that is, to put the glucose, amino acids, some salts and water back into the blood stream. The tubule therefore separates the waste products from the useful components of the blood.

Reabsorption is *selective*. The amount of water and salts taken back into the blood stream is carefully adjusted and monitored.

The filtrate passes from the renal capsule and into the first convoluted tubule. Look at Fig. 8.12. Notice that *all* the glucose and *all* the amino acids are reabsorbed back into the blood stream in this tubule. The water moves back into the blood stream by diffusion. Glucose and amino acids move back into the blood stream both by diffusion and by active transport.

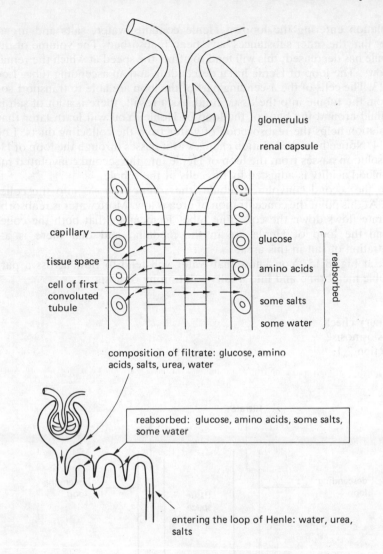

composition of filtrate: glucose, amino
acids, salts, urea, water

reabsorbed: glucose, amino acids, some salts,
some water

entering the loop of Henle: water, urea,
salts

Figure 8.12   Reabsorption in the first convoluted tubule

■ Memory check
What is diffusion?
What is active transport?
See Section 2.1.

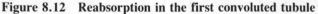

─ **Exercise** ─

1 Why is diffusion alone an inadequate way of moving glucose and amino
   acids back into the blood stream?
2 Energy is needed for active transport to take place in the cells of the first
   convoluted tubule. Where do these cells get their energy from?

The solution entering the loop of Henle contains water, salts and urea. Remember that the other substances have been reabsorbed. The volume of fluid in the tubule has decreased; this will tend to affect the speed at which the remaining fluid flows. The loop of Henle has a descending and an ascending tube. Look at Fig. 8.13. The cells of the ascending part of the loop are able to transport sodium salts from the tubule into the tissue fluid. As a result, there is a lot of salt in the tissue fluid around the base of the loop of Henle. You will learn later that this salty solution helps the reabsorption of water from the collecting ducts. Look at Fig. 8.14. Notice how the solution changes as it passes through the loop of Henle.

The solution passes from the loop of Henle into the second convoluted tubule, where blood acidity is adjusted by the cells of the tubule.

From the second convoluted tubule, the solution passes into the collecting tubule. At this point the concentration of urea is low. More water is reabsorbed as the filtrate flows down the collecting tube. Remember that both the collecting duct and the loop of Henle are in the medulla, and that there is a high concentration of salt in this area.

Look at Fig. 8.15. You know that water will tend to flow across a partially permeable membrane and into a salty solution by osmosis.

■ **Memory check**
What is osmosis?
See Section 2.1.

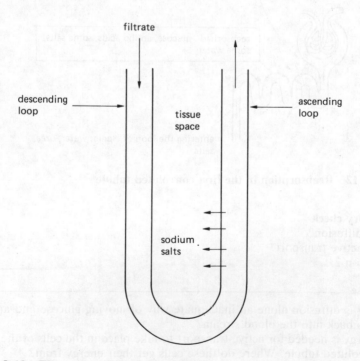

Figure 8.13   The action of the loop of Henle

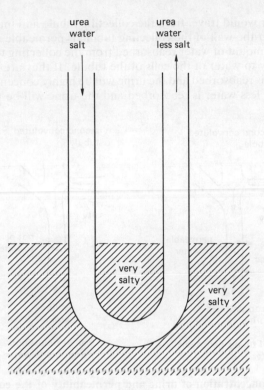

Figure 8.14   The action of the loop of Henle

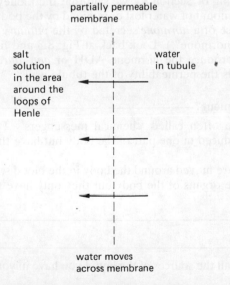

Figure 8.15   Osmosis (movement of water across a partially permeable membrane) in the loop of Henle

Therefore water would travel from the collecting tubule and into the tissue space of the medulla *if* the wall of the collecting tubule is permeable to water. Look at Fig. 8.16. The amount of water reabsorbed from the collecting tubule depends on the permeability to water of the cells of the tubule. If they are very permeable a lot of the water is reabsorbed and the urine will be highly concentrated. If they are less permeable, less water is reabsorbed and the urine will be more dilute.

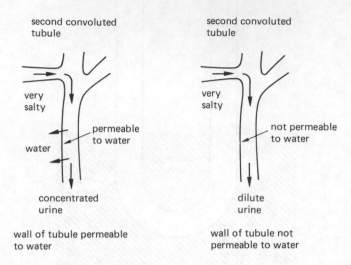

Figure 8.16   Concentration of urine and permeability of the collecting tube wall

At the beginning of Section 8.2 we said that the kidneys were responsible for controlling the amount of water lost or retained by the body. The kidneys are able to do this because of a *hormone* secreted by the *pituitary gland* (read the boxed note 'What is a hormone?'). Look back at Fig. 8.3 and find the pituitary gland. The gland can produce the hormone ADH or *anti-diuretic hormone*, and this hormone controls the permeability of the tubule walls.

What is a hormone?

Hormones are often called 'chemical messengers'. They are compounds which are produced in one part of the body but have their effect in another part.

Hormones are moved around the body in the blood stream. They will pass through all the organs of the body but they only have their effect on *target organs*.

Exercise

Make a list of all the sources of the liquid you have in your body. Call this list A.

Make a list of all the ways in which water is lost from your body. Call this list B.

1 How would the volume of liquids in lists A and B change if you were physically active in a hot dry atmosphere for most of the day?
2 How would the volume of liquids in lists A and B change if you were physically inactive in a hot dry atmosphere for most of the day?

Summary

■ Blood undergoes *ultrafiltration* in the renal capsule.

■ *Selective reabsorption* of glucose, amino acids and some salts and water occurs in the first convoluted tubule.

■ The acidity of the blood is regulated in the second convoluted tubule.

■ Water reabsorption is controlled in the collecting tubule.

■ The control of water reabsorption is due to a hormone, ADH, secreted by the pituitary gland.

Questions on Section 8.3

1 Look at the table below:

Substance	Blood	Filtrate in renal capsule	Urine
glucose			
amino acids			
urea			
protein			
water			
salts			

Put a tick (√) if the substance is present in the fluids listed at the top of the table or a cross (×) if it is not present.
2 Do you think humidity of the air has any effect on the volume of urine produced? Explain your answer.

8.4 Kidney failure, dialysis and transplants

Some diseases cause permanent damage to the tissues of the kidney. When the kidney fails to do its job properly the body gradually becomes poisoned by the increasing amounts of urea in the blood. The urea has to be removed, or the body will die. Urea can be removed by dialysis. The alternative to dialysis is replacement of the diseased kidney by a transplanted one.

Dialysis

Dialysis is a way of removing soluble substances from a liquid (in this case, toxic substances from blood) by diffusion across a partially permeable membrane.

■ **Memory check**
What is diffusion?
See Section 2.1.
Note: Do not go on until you are sure that you understand diffusion.

There are two types of dialysis:

● haemodialysis
● CAPD (continuous ambulatory peritoneal dialysis).

Let's look at *haemodialysis* first. In haemodialysis:

● blood is taken out of the body
● it is treated with an anticoagulant so that it will not clot when it touches the tubes and membranes outside the body
● it is passed over a partially permeable membrane and then back into the body.

Look at Fig. 8.17.

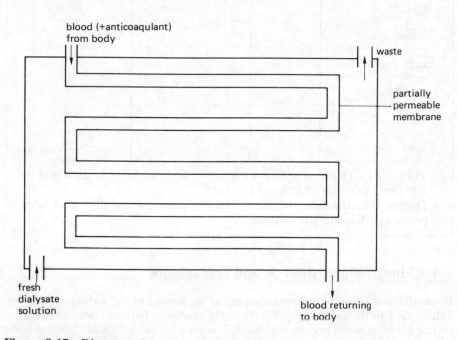

Figure 8.17   Diagram of a haemodialysis circuit

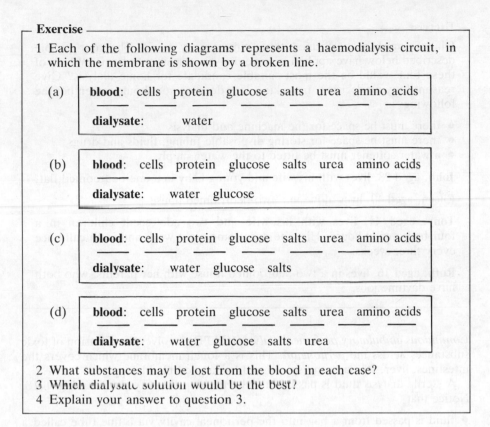

Exercise

1 Each of the following diagrams represents a haemodialysis circuit, in which the membrane is shown by a broken line.

(a)
> **blood**: cells protein glucose salts urea amino acids
>
> **dialysate**: water

(b)
> **blood**: cells protein glucose salts urea amino acids
>
> **dialysate**: water glucose

(c)
> **blood**: cells protein glucose salts urea amino acids
>
> **dialysate**: water glucose salts

(d)
> **blood**: cells protein glucose salts urea amino acids
>
> **dialysate**: water glucose salts urea

2 What substances may be lost from the blood in each case?
3 Which dialysate solution would be the best?
4 Explain your answer to question 3.

The blood loses toxic substances in the dialysis and then returns to the tissues, where it can pick up more accumulated toxins from the tissue fluids. Toxins in cells are then able to pass into the tissue fluid. The blood thus needs to be passed through the dialyser several times to ensure the complete removal of toxins from the body.

Let us consider the design of dialysers for a moment. There are three main types:

- in a hollow fibre dialyser, blood is passed through thousands of tiny tubes surrounded by the dialysate
- in a coil dialyser, blood is passed into a long flat bag which is coiled and surrounded by dialysate
- in a flat bed dialyser, blood is passed between stacked membranes.

In each case a large surface area is available for diffusion, and that the dialysate and blood always flow in opposite directions. The dialysate always has a lower concentration of urea and salts than the blood from which it is separated by the membrane. Diffusion from the blood is therefore efficient and fast.

Look at Fig. 8.17 once more. The pressure of the dialysate can be changed so as to adjust the amount of water to be lost from the blood.

Exercise

Dialysis must be performed three times every week. The four people described below have applied to have dialysis machines at home. Which of these four would be the most suitable candidate for home dialysis? Give reasons for your decision. In making your decision you should remember the following:

- there must be space for the machine and dialysis
- there must be space for storing disposable tubing, fluids and drugs
- a water softener must be fitted to the water supply.

John, aged 28, lives with his wife and young baby in a one-bedroomed flat.

Eileen, aged 40, lives alone in a two-bedroomed house.

Tony, aged 44, lives with his wife and two adolescent children in a four-bedroomed house. His wife does some night work and is on call once every three weeks.

Ruth, aged 16, lives in a two-bedroomed house with her parents, who both have daytime jobs.

Continuous ambulatory peritoneal dialysis (CAPD) involves the diffusion of toxic substances across the *peritoneum*. This is a tough membrane which covers the intestines, liver, stomach and spleen and has a good blood supply.

A sterile dialysis fluid is passed into the peritoneal cavity. Look at Fig. 8.18. Notice that

- fluid is passed from a bag into the peritoneal cavity via a fine tube called a catheter
- fluid can be drawn off later and collected in the bag X.

Four or five times a day, two litres of fluid are run into the peritoneum. The diffusion goes on continuously, so the patient does not have to be connected up to a machine for hours at time, as in dialysis.

Exercise

What are the advantages of CAPD over haemodialysis?

The peritoneum can easily become infected, however. The patient must be very careful to use sterile fluid, and must attend the hospital once a month to have the external connector of the catheter changed.

Exercise

Fluid dialysate in haemodialysis does not have to be sterile. Why is this?

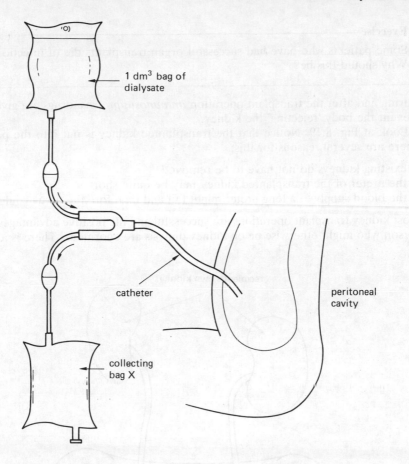

Figure 8.18   CAPD circuit

Kidney transplant

Exercise

What are the advantages of a kidney transplant over dialysis?

Transplanting a kidney into the body of another person is more complicated than transplanting a seedling!

There are cells wandering around your body examining the surfaces of all the other cells to see if they are your own – that is, 'self'. If a cell is not 'self' it is destroyed. This is known as the *immune response* and can result in rejection of a transplanted organ. So if an organ is to be transplanted successfully the body must *either* be prevented from recognising 'non-self', *or* be prevented from destroyed the 'non-self' – and it is possible to do the latter with drugs.

Exercise

Some patients who have had successful organ transplants die of infections. Why should this be?

During and after the transplant operation *immunosuppressive* drugs are given to prevent the body 'rejecting' the kidney.

Look at Fig. 8.19. Notice that the transplanted kidney is put into the pelvis. There are several reasons for this:

- existing kidneys do not have to be removed
- the ureter of the transplanted kidney may be quite short
- the blood supply to a long ureter might fail and therefore the tissue could die.

Most kidney transplant operations are successful and the positive advantages to a person who might otherwise be on kidney dialysis are enormous. There is a long

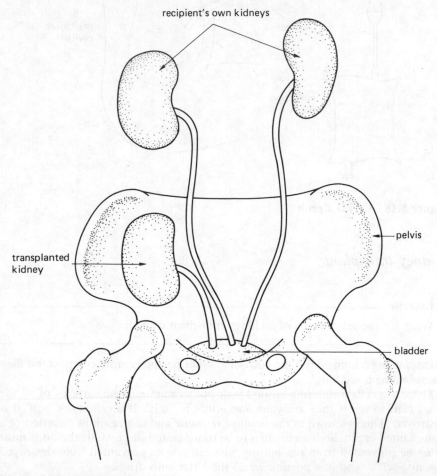

Figure 8.19   Kidney transplant

waiting list for the operation, largely due to the lack of suitable kidneys. Most transplanted kidneys are taken from young or middle-aged people who have had an accident resulting in severe and irreparable brain damage but whose breathing is maintained on a ventilator. (The oxygen keeps the kidneys functioning.) If such a person is carrying a kidney donor card, his or her kidneys can be made available to a patient with kidney failure.

--- Summary ---

■ Kidney failure results in a build-up of toxic substances in the body.

■ *Dialysis* separates toxic substances from the blood.

■ In dialysis, toxic substances pass across a partially permeable membrane and into a dialysate by diffusion.

■ *CAPD* involves the diffusion of toxic substances across the wall of the perineum into a fluid which has been passed into the peritoneal cavity.

■ *Kidney transplants* are preferable to dialysis.

■ Rejection of a transplanted organ is prevented by the use of *immuno-suppressive drugs*.

--- PROJECT WORK ---

1 What other organs can be transplanted?
2 Find out what is meant by *tissue typing*. Why is this important when considering organ transplants?

9 **Support and movement**

9.1 **The human form**

The skeleton of human beings is inside the body. It protects many vital structures like the brain, heart, lungs, the large blood vessels and sense organs of sight, smell and hearing. It forms a rigid framework to which muscles are attached, and therefore allows movement. Many bones of the body act as levers, so that the maximum amount of work is done for the minimum effort. Red and white blood cells are made in the marrow of many bones. Bones contain large quantities of calcium which can be passed into the blood stream if the body requires it.

Human beings are part of a sub-phylum of animals called the *vertebrata*. All vertebrates have a vertebral column (backbone) which encloses a spinal cord.

Summary

■ The functions of the skeleton are:

to give shape to the body
to give support
to give protection
to give surfaces for muscle attachment.

■ Red and white blood cells are made in the bone marrow.

Questions on Section 9.1

1 Name three of the functions of the skeleton.
2 What characteristics do you think bone should have if it is to fulfil all the functions described in Section 9.1?
3 'Human beings are vertebrates'. What does this statement mean?

9.2 **The nature and structure of bone**

Look at Fig. 9.1. Notice that:

● mature bone has a dense outer layer of *compact bone*
● under the compact bone and especially at the end of the bone there is *spongy bone* (sometimes called *cancellous bone*)
● the shaft (the long middle section of the bone) is filled with bone marrow

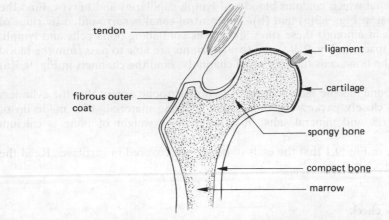

Figure 9.1   Section through the head of the femur (mature bone)

- because the interior of the bone is less dense, the bone is light and strong; spongy bone is concentrated in areas that need to be particularly strong
- there is a layer of *cartilage* at the end of the bone to protect it, cushion it and allow easy movement
- bones are held to one another by *ligaments*. These are slightly elastic, to allow some 'give' when joints are stretched or are under stress.

Closer examination of bone reveals that it is made up of many closely packed cylinders. Look at Fig. 9.2. Under a microscope you can see that each cylinder has

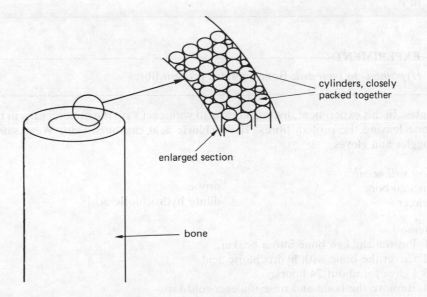

Figure 9.2   Structure of bone (diagrammatic)

a central canal which contains blood and lymph capillaries and nerves. Find the central canal in Fig. 9.3(a) and (b). The central canal is surrounded by rings of bony plates in amongst these rings are spaces containing bone cells and lymph. Find these spaces in Fig. 9.3(a) and (b). Nutrients are able to pass from the blood stream to the bone cells through small channels. Find the channels in Fig. 9.3(a) and (b).

Spongy bone has the same structure as compact bone except that the cylinders are not so closely packed together. Bone is living material. It is made up of protein fibres and mineral salts: 70% of the dry weight of bone is calcium phosphate.

You saw in Fig. 9.1 that the ends of bones are covered in cartilage. Read the boxed note 'Cartilage'.

■ **Memory check**
What mineral salts do you need to make strong bones?
What vitamin do you need to make strong bones?
See Section 5.1.

Cartilage

Cartilage is an important tissue. Many bones begin in the foetus as cartilage which later develops into bone. Cartilage forms discs or pads between the vertebrae, part of the larynx, trachea, bronchi and the pinna of the ear. Cartilage is a smooth, bluish-white tissue which is strong and slightly flexible. An especially strong and flexible cartilage is needed between the vertebrae and in the knee and shoulder joints, so the cartilage in these areas has fibres in it.

EXPERIMENT

Hypothesis to be tested: Bone contains protein fibres

Notes: In this experiment, hydrochloric acid will react with the mineral salts in the bone leaving the protein fibres. **Hydrochloric acid can burn skin. Wear safety goggles and gloves**.

You will need:
chicken bone   probe
beaker   dilute hydrochloric acid

Method
1 Put the chicken bone into a beaker.
2 Cover the bone with hydrochloric acid.
3 Leave for about 24 hours.
4 Remove the bone and rinse under a cold tap.
5 Examine the bone. Try to bend it. Press the bone with the end of the handle of the probe.

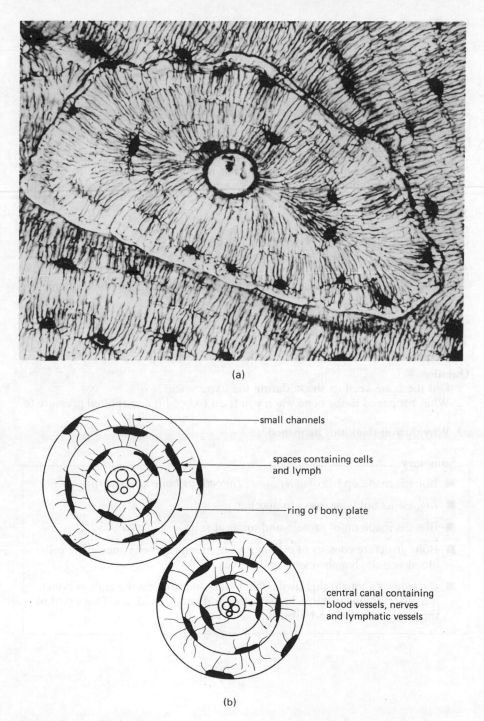

(a)

(b)

Figure 9.3 **(a) Section of a bone; (b) drawing showing the detail of the cylinders**

Questions
1 Did the bone keep its original shape during the experiment?
2 What happened to the bone when you tried to bend it and applied pressure to it?
3 Why do you think this happened?

┌─ **EXPERIMENT** ───
│ *Hypothesis to be tested:* Bone contains mineral salts
└──

Notes: This experiment produces smells that may be uncomfortable to work in and may cause irritation to the breathing system. Use a fume cupboard if possible.

You will need:
chicken bone   probe
tongs   bunsen burner

Method
1 Wearing eye protection, use the tongs to hold the bone in the bunsen flame for about two minutes.
2 Allow the bone to cool.
3 Examine the bone. Try to bend it. Press the heated end of the bone with the handle of the probe.

Questions
1 Did the bone keep its shape during the experiment?
2 What happened to the bone when you tried to bend it and applied pressure to it?
3 Why do you think this happened?

┌─ **Summary** ──
│ ■ Bone is made up of *compact bone*, *cancellous bone* and *marrow*.
│
│ ■ *Ligaments* hold one bone to another.
│
│ ■ Bone is made up of protein and mineral salts.
│
│ ■ Bone structure consists of cylinders containing rings of bone, living cells, blood vessels, lymph vessels and nerves.
│
│ ■ *Cartilage* is a tough, slightly flexible tissue that protects the ends of bones, is a model for many bones in the foetus and young child, and forms part of the larynx, trachea and bronchi.
└──

Questions on Section 9.2

Look at Fig. 9.4.

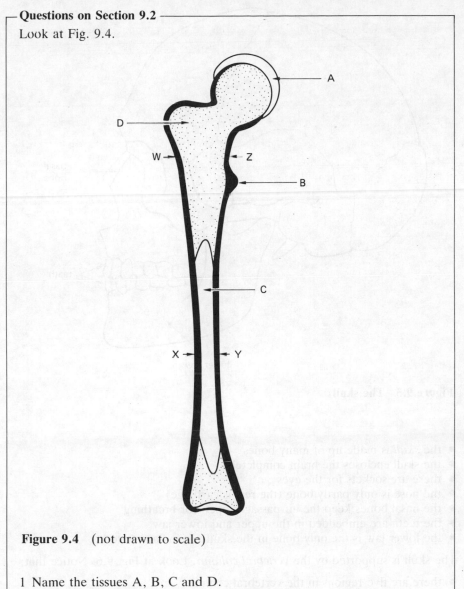

Figure 9.4 (not drawn to scale)

1 Name the tissues A, B, C and D.
2 What is the function of the tissue labelled A?
3 What is the function of the tissue labelled D?
4 Draw transverse sections through the points labelled W–Z and X–Y.

9.3 Bones, bones and more bones!

In this section you are going to get to know some of the bones of the body. Let's
start with the head. Look at Fig. 9.5. Notice that:

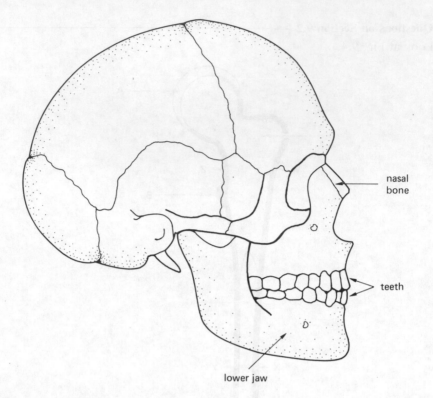

Figure 9.5   The skull

- the *skull* is made up of many bones
- the skull encloses the brain completely
- there are sockets for the eyes
- the nose is only partly bone (the rest is cartilage)
- the nasal bones keep the air passages open for breathing
- the teeth are embedded in the upper and lower jaw
- the lower jaw is the only bone in the skull which can move.

The skull is supported by the *vertebral column*. Look at Fig. 9.6. Notice that:

- there are five regions in the vertebral column
- the column is naturally curved
- the *sacrum* and *coccyx* are made up of several bones fused together.

Let's look at a thoracic vertebra in a little more detail. Look at Fig. 9.7. Notice:

- the *body*, which becomes larger towards the load-bearing region (refer back to Fig. 9.6). Each vertebral 'body' must bear the weight of the rest of the whole body above it
- the *vertebral canal*, through which the spinal cord passes
- the processes which make joints with the ribs; these joints move during breathing.

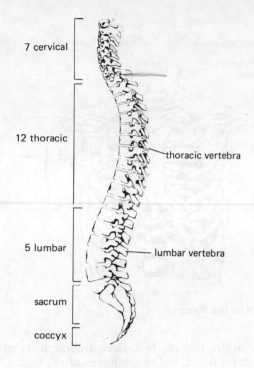

7 cervical

12 thoracic

thoracic vertebra

5 lumbar

lumbar vertebra

sacrum

coccyx

Figure 9.6   The vertebral column (lateral view)

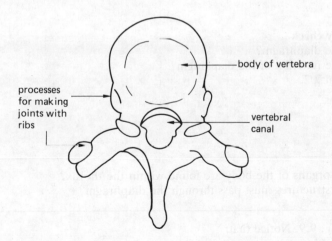

body of vertebra

processes
for making
joints with
ribs

vertebral
canal

Figure 9.7   A thoracic vertebra

■ **Memory check**
Which muscles pull the ribs upwards during breathing?
How does the movement of the ribs help in breathing in?
See Section 7.2.

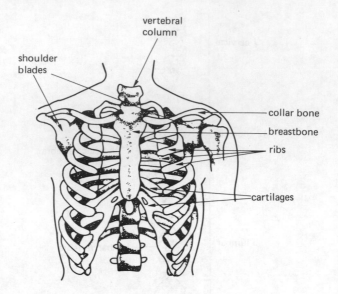

Figure 9.8   Bones of the thorax

Look at Fig. 9.8. Notice that the thoracic vertebrae form part of the chest or *thorax*. The diaphragm is attached to the bones of the thorax.

■ **Memory check**
What is the diaphragm?
What is the function of the diaphragm?
See Section 7.2.

┌─ **Exercise** ──┐
│   │
│   1 What organs of the body are found within the thorax?   │
│   2 What structures must pass through the diaphragm?   │
│   │
└───┘

Look at Fig. 9.9. Notice that:

• the shoulder girdle is made up of the *collar bones* and *shoulder blades*
• the arm is made up of the *humerus*, *radius*, *ulna* and *hand bones*
• the small bones of the hand make up the wrist and the longer bones make up the hand and fingers.

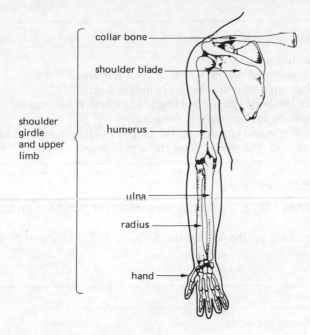

Figure 9.9 Bones of the shoulder girdle

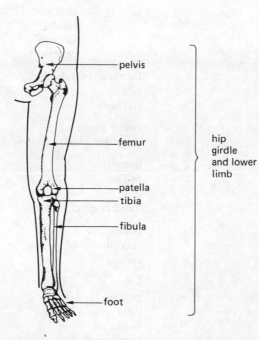

Figure 9.10 Bones of the hip girdle and leg

Exercise

This is a short practical exercise.

1 Bend your left arm.
2 Put your right hand on the elbow of your left arm.
3 Rotate your wrist as if you were opening a door.
4 Notice that the bone you are holding at the elbow is not moving. The bone you are holding at the elbow is the *ulna*.
5 The bone that is rotating around the elbow and helping the hand to move is the *radius*. So, the *radius* helps the wrist *rotate*.

Look at Fig. 9.10. Notice that:

- the pelvis forms a basin, supporting the contents of the abdomen and the rest of the body
- the bones making up the leg are the *femur*, *patella*, *tibia* and *fibula*, and the foot bones.

Summary

Part of the body	Bones
head	many skull bones
vertebral column	five areas: cervical
	thoracic
	lumbar
	sacrum
	coccyx
thorax (chest)	thoracic vertebrae
	ribs
	sternum
shoulder girdle	collar bone
	shoulder blade
arm	humerus
	radius
	ulna
	hand bones
hip girdle	pelvis
	sacrum
	coccyx
leg	femur
	patella
	tibia
	fibula
	foot bones

Questions on Section 9.3

1 Look at Fig. 9.11. Label the bones.

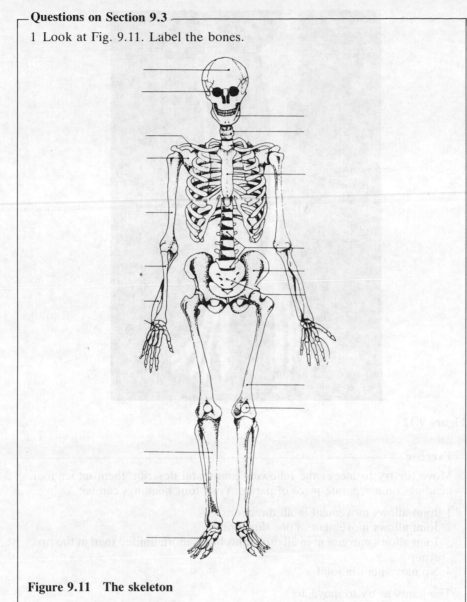

Figure 9.11 The skeleton

2 Look at Fig. 9.12.
Where are these bones to be found in the body?
What are the names of the bones in the photograph?

9.4 **Joints**

A joint is formed when two or more bones meet.

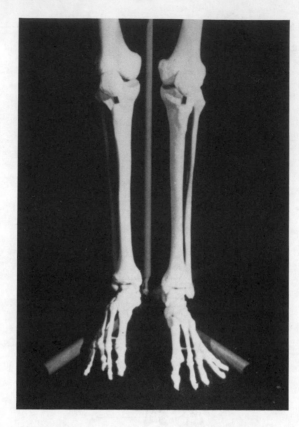

Figure 9.12

Exercise

Move (or try to move) the following joints and describe them under four headings on a separate piece of paper. Your four headings can be:

1 Joint allows movement in all directions
2 Joint allows movement in one direction
3 Joint allows movement in all directions but is more limited than in the first group
4 No movement in joint

The joints to try to move are:

fingers	ribs articulating with the sternum
thumb	ribs articulating with the vertebrae
elbow	femur/pelvis
shoulder/arm	knee
neck	teeth in their sockets
vertebral column	

Try to remember the names of the bones involved as you think of the joints.

In doing this exercise you will have found that joints can give different types of movement. Joints are classified according to how much movement they give. We are going to look at the different types of joint now.

Immovable joints

These joints can also be called *fibrous* or *fixed joints*. These joints give no movement at all. Look at Fig. 9.13. Notice that there is fibrous tissue between the two bones. The joints between the bones of the skull are of this type. So are the joints formed by the teeth in their sockets.

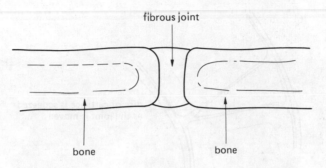

fibrous joint

bone   bone

Figure 9.13 A fibrous joint (diagrammatic)

Slightly movable joints

These joints can also be called *cartilaginous joints*. These joints give limited movement. Look at Fig. 9.14(a). Notice that there is a disc of cartilage between the two bones. When this joint moves the disc of cartilage is squeezed. Look at Fig. 9.14(b). The joints between the vertebrae are of this type.

Freely movable joints

These joints can also be called *synovial joints*. These joints give easy movement between the bones. There are two main types of synovial joint. The type is dependent on the amount of movement the joint gives. Look at your responses to the exercise at the beginning of this section. The joints under your first two headings are freely movable joints, the first set giving you movement in all directions and the second movement in one direction only.

Joints which give you movement in all directions are ball-and-socket joints. Look at Fig. 9.15.

Joints which give you movement in one direction are hinge joints. Look at Fig. 9.16.

Ball-and-socket and hinge joints have the same basic structure. They are all surrounded by a *synovial membrane* which secretes a lubricating *synovial fluid*. Look at Fig. 9.17. Notice that the ends of the bones are covered in cartilage, which is completely enclosed by the synovial membrane.

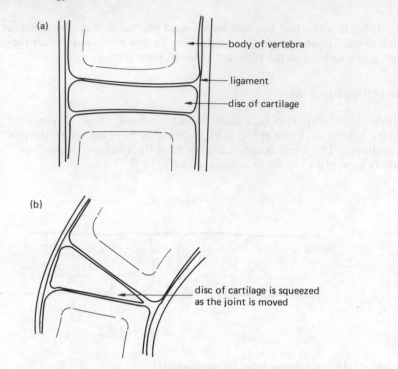

Figure 9.14 (a) A cartilaginous joint (diagrammatic); (b) movement in a cartilaginous joint

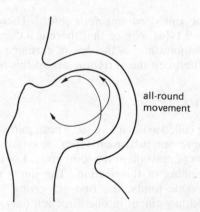

Figure 9.15 A ball-and-socket joint (diagrammatic)

■ **Memory check**
Why is there cartilage at the ends of bones at joints?
See Section 9.2.

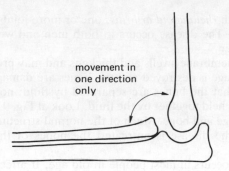

movement in
one direction
only

Figure 9.16   A hinge joint (diagrammatic)

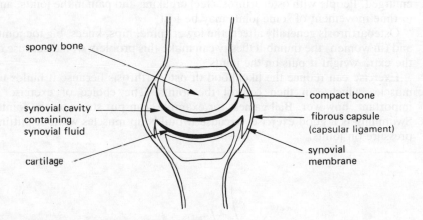

spongy bone

synovial cavity
containing
synovial fluid

cartilage

compact bone

fibrous capsule
(capsular ligament)

synovial
membrane

Figure 9.17   Section of a synovial joint

Synovial fluid is a thick sticky substance, like egg white. It lubricates the joint allowing free movement of the bones. Fig. 9.18 is a simplified diagram of a joint, allowing you to see all the structures. In reality, however, the bones touch each other; they are held together by the synovial fluid in much the same way as two pieces of glass can be held together by a film of water.

Look at Fig. 9.17 again. Notice that the capsular ligament encloses the joint. It allows movement but protects the joint from injury. Other ligaments go across the joint, holding the bones together. (Remember that ligaments hold bone to bone.) These are not shown in Fig. 9.17 because this is a section. Look at Fig. 9.18 to see how ligaments hold bone to bone around a joint.

The elbow, knee, shoulder, hip and finger joints are all freely movable joints.

Problems with joints

One of the commonest problems associated with joints is *arthritis*. This word describes an inflammation of a joint and its surrounding structures. There are two types of arthritis: *rheumatoid arthritis* and *osteoarthrosis* (which used to be called osteoarthritis).

In a person with *rheumatoid arthritis*, one or more joints become painful, stiff and swollen. The disease occurs in both men and women but is more common in women.

The synovial membrane swells and thickens and may produce extra fluid. In time the cartilage is destroyed and the bones are damaged. Look at Fig. 9.19(a). Notice that the bones are separated by fluid; normally the bones touch each other, held together by the fluid. Look at Fig. 9.19(b). Notice the damage to cartilage and bone. Much of the normal structure of the joint has disappeared. If the disease is untreated the bones of the joint may fuse together.

Osteoarthrosis occurs in most people in old age. It affects weight-bearing cartilage surfaces. The cartilage becomes thin and finally disappears as cells are not replaced. The bone under the cartilage may become thickened and enlarged. People with osteoarthrosis feel creaking and pain in the joints, and in time movement of some joints may be lost.

Osteoarthrosis generally affects the lower spine, hips, knees, big toe joints and (in women) the thumb. Obesity can make this problem worse because of the extra weight it puts on the joints.

Exercise can reduce the likelihood of osteoarthrosis because it builds up muscles which can then control the joints. The choice of exercise is important, however. Ball games, for example, can put stress on the joints. Swimming is a good exercise, because it builds up muscles without putting pressure on joints.

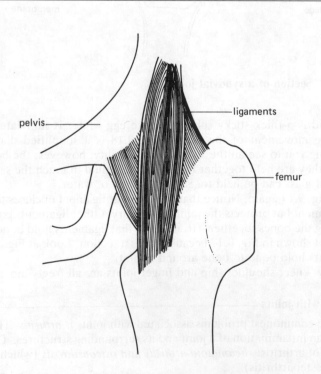

Figure 9.18 Ligaments holding the femur and pelvis together

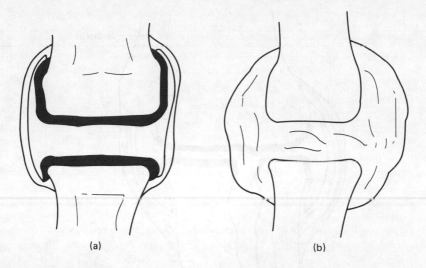

Figure 9.19 **Rheumatoid arthritis: (a) early stages, (b) advanced stage**

Summary

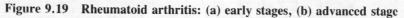

Joint	Other names	Example
immovable	fibrous	skull teeth in sockets
slightly movable	cartilaginous	vertebrae
freely movable	synovial	elbow shoulder

■ Rheumatoid arthritis is a painful swelling of the joint caused by thickening of the synovial membrane and the production of extra fluid.

■ Osteoarthrosis affects weight-bearing joints and results in degeneration of the joint.

Questions on Section 9.4

Look at Fig. 9.20.

1 Is this joint freely movable or slightly movable?
2 What sort of movement can this joint give?
3 Label the synovial fluid, synovial membrane, capsular ligament and cartilage.
4 What are the functions of the structures you have labelled?

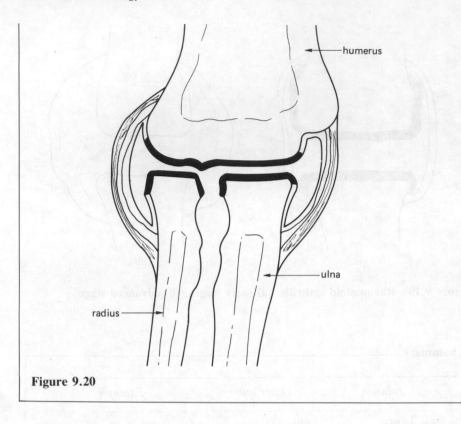

humerus

ulna

radius

Figure 9.20

9.5 Muscles and how they work

There are three types of muscle:

- voluntary muscle — skeletal
- involuntary muscle — smooth
- cardiac muscle.

Muscles bring about movement. They can *contract* and *relax*.

Voluntary muscles can be stimulated to contract and cause movement under conscious control. That is, if you want to move your arm, your brain will send impulses via effector neurones to your arm muscles, causing them to contract. Look at Fig. 9.21(a). Notice that muscle is made up of bundles of striped fibres containing nuclei, but you cannot relate the nuclei to individual cells. Look at Fig. 9.21(b). Find the nuclei and the stripes.

Contraction of *involuntary muscle* is not under conscious control. Involuntary muscle is found in the walls of blood and lymph vessels, the digestive system, diaphragm, uterus and bladder. Look at Fig. 9.22. Notice that this muscle is made up of distinct spindle-shaped cells. Each cell has its own nucleus. The cells form a sheet of muscle.

The heart is made up of *cardiac muscle*. Look at Fig. 9.23. Notice that the muscle has stripes which are a little like voluntary muscle. The contraction of this

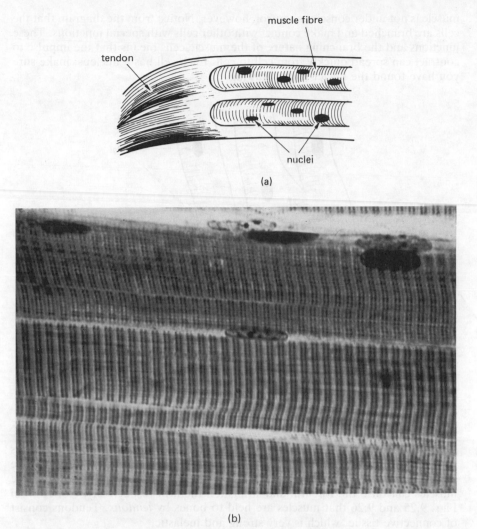

muscle fibre

tendon

nuclei

(a)

(b)

Figure 9.21   Voluntary muscle

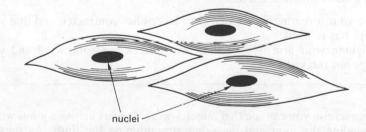

nuclei

Figure 9.22   Cells of involuntary muscle

muscle is not under conscious control, however. Notice from the diagram that the cells are branched and make contact with other cells with special junctions. These junctions and the branching nature of the muscle cells means that the impulse to contract can spread quickly from cell to cell. Each cell has a nucleus; make sure you have found the nuclei in the drawing.

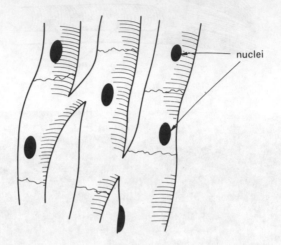

Figure 9.23   Cardiac muscle

How muscles work

Look at Fig. 9.24. Find the neurone and the motor end plates. Muscle is stimulated to contract by impulses from these motor end plates. Consider the muscles of the arm. Look at Fig. 9.25. Find the biceps and triceps muscles. These muscles can raise the forearm and straighten the arm respectively. Notice from Figs. 9.25 and 9.26 that muscles are held to bones by *tendons*. Tendons consist of connective tissue, which is very strong and inelastic.

┌─ Exercise ──────────────────────────────

This is a short practical exercise.

1 Raise your forearm. Notice that your biceps has contracted and that your triceps has relaxed.
2 Straighten your arm. Notice that your triceps has contracted and your biceps has relaxed.

From this exercise you can see that muscles work in pairs across a joint, with one muscle bending the joint and the other straightening the limb. As they have opposite effects on the limb they are said to be *antagonistic*.

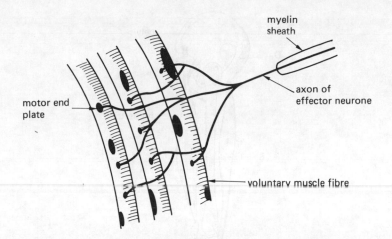

myelin
sheath

axon of
effector neurone

motor end
plate

voluntary muscle fibre

Figure 9.24   How a stimulus reaches muscle fibres

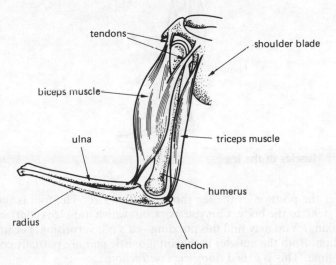

tendons

shoulder blade

biceps muscle

triceps muscle

ulna

humerus

radius

tendon

Figure 9.25   Muscles of the arm

Exercise

Look at Fig. 9.26.
1 Which muscle must contract to raise the foot (Fig. 9.27(a))?
2 Which muscle must contract to bring the heel up and the foot back (Fig. 9.27(b))?
3 Which muscle must contract to bring the upper leg back (Fig. 9.27(c))?
4 Which muscle must contract to bring the upper leg forwards (Fig. 9.27(d))?

If you have difficulty with this exercise move your own leg into the positions described and feel which muscles are contracting.

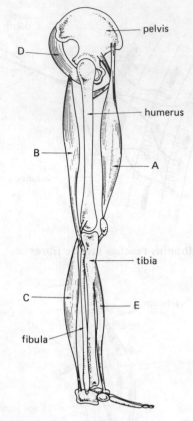

Figure 9.26   Muscles of the leg

Now consider the position of the leg shown in Fig. 9.26. The foot is flat on the ground and at 90° to the body. Can you work out which muscle is contracting and which is relaxing? You may find this puzzling – it's not surprising, because this is a trick question! Both the muscles in the antagonistic pair are partially contracted at the same time. This is called *isometric contraction*.

Levers and limbs

When a muscle contracts it gets shorter and thicker. This shortening pulls the bone to which it is attached. Limb muscles cause movement across joints and work the bones as levers. Look at the diagram below.

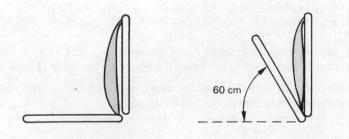

60 cm

Notice that a muscle contracts a short distance (the biceps shortens only by 0.8 cm), but its attachment is close to a joint, so the movement at the end of the limb is magnified (the wrist end of the forearm moves 60 cm).

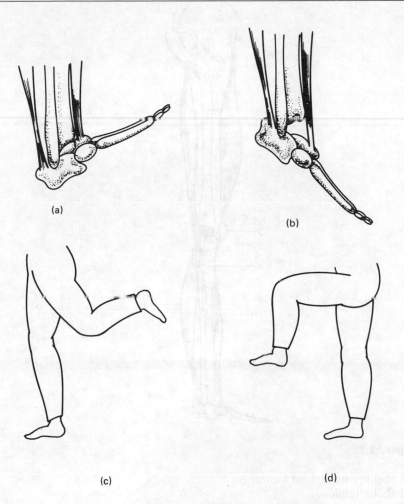

(a)

(b)

(c)

(d)

Figure 9.27 Muscle contraction results in different leg positions

Summary

■ There are *voluntary*, *involuntary* and *cardiac* muscles.

■ Muscles can contract and relax.

■ Muscles work in pairs. The pairs are said to be *antagonistic*.

■ Simultaneous partial contraction in a pair of antagonistic muscles is called *isometric contraction*.

Questions on Section 9.5

Look at Fig. 9.28.

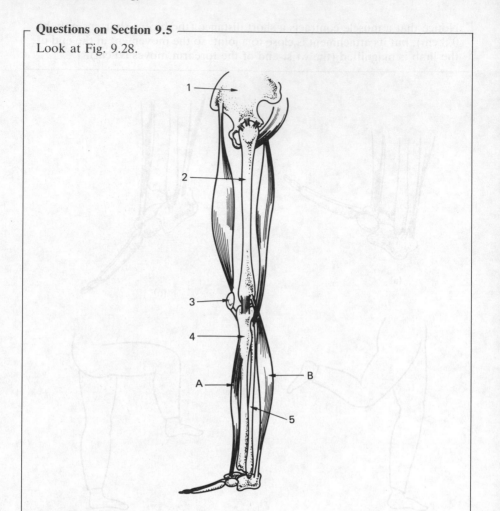

Figure 9.28

1 Label the numbered bones.
2 Label a tendon.
3 Describe two properties of a tendon.
4 Label a ligament.
5 Describe two properties of a ligament.
6 Explain what will happen to the foot when muscle A contracts.
7 Explain what will happen to the foot when muscle B contracts.

9.6 Feet

For the first 18 years of our lives our feet are growing. If growing feet are under pressure, bones and joints become distorted leading to deformity and pain. Look

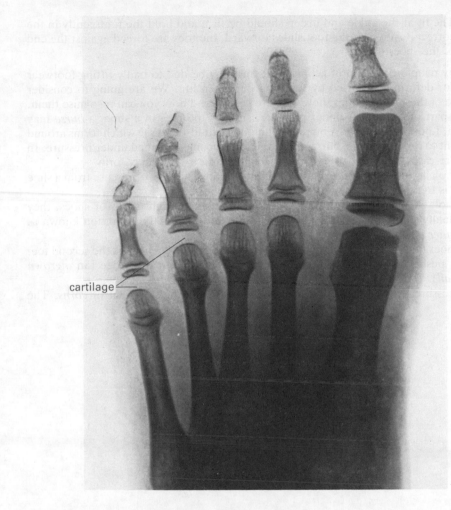

cartilage

Figure 9.29 X-ray photograph of a child's foot

at Fig. 9.29. Notice the areas of cartilage. These areas will gradually develop into bone. Cartilage has some flexibility and it is therefore easy to distort. Distorted cartilage will give rise to distorted bone. So what can distort these developing bones?

- socks that are too small or too narrow and tight
- tight all-in-one stretch-knit baby suits that squash the toes
- shoes that are not the correct width or length
- shoes that fail to give support to the foot.

What qualities should a good shoe have?

1 It should hold the foot firm while allowing it plenty of room. Its side walls should be straight, to give depth; this will avoid pressure from the sides.

2 The fit at the ankle and instep should be firm and hold the foot gently in the correct position. If the foot slides forward, the toes are forced against the end of the shoe.

Many people complain of aching feet; this may be due to badly fitting footwear and/or deformities caused by footwear earlier in life. We are going to consider some of these deformities, how they are caused and how you can recognise them.

If part of the foot is under pressure, such as the toes in a shoe, a *bursa* may form. Look at Fig. 9.30. A bursa or bunion is a fluid-filled sac which forms around a joint to protect it. The bursa may swell and become inflamed under pressure. In severe cases, an extra growth of bone called an *exostosis* may form.

Look at Fig. 9.31. A *bursitis* is a painful swelling caused by pressure from a shoe that is too tight or too loose and continually rubs.

Look at Fig. 9.32. If toes are cramped together by tight socks or shoes, they gradually come to overlap and become crooked. This is the condition known as *hammer toes*.

Shoes that are too narrow can cause the big toes to press against the second toe. This pushes up a ridge of flesh. The toenail grows into this ridge (an *ingrown toenail*).

Pressure from ill-fitting shoes can also cause the development of *corns*. The

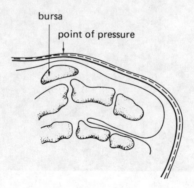

Figure 9.30   A bursa or bunion

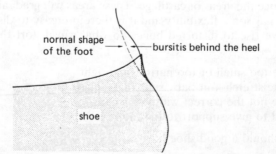

Figure 9.31   A bursitis

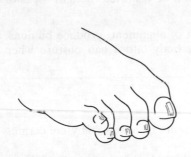

Figure 9.32   Hammer toes   **Figure 9.33   High-heeled shoes**

skin in the area under pressure becomes thick and hard: the thicker the skin, the more painful the corn becomes.

The foot forms an arch which supports the body. The arch itself is supported by the heel, the base of the big toe and the bones of the little toe. If the arch is improperly supported – as it is when high-heeled shoes are worn – it may collapse (a *fallen arch*).

Exercise

Some women like to wear narrow-toed, high-heeled shoes. Look at Fig. 9.33.

1 Why might wearing this type of shoe cause the development of a bursa, hammer toes and corns?

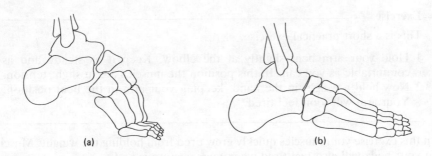

(a)   (b)

Figure 9.34   Alignment of foot bones (a) when wearing high-heeled shoes, (b) when wearing flat shoes

2 Compare Fig. 9.34(a) and (b).
  How are the high-heeled shoes changing the normal alignment of the bones of the feet?
3 Someone who wears high-heeled shoes for several hours may develop backache. Why do you think this happens?

Summary

■ The bones in the feet are not fully formed until the age of 18.

■ The cartilage in the developing bones can be distorted by tight socks or baby suits and ill-fitting shoes.

■ Poorly fitting shoes can push bones out of alignment, produce bunions, hammer toes and corns, and force the body into a bad posture when standing.

Questions on Section 9.6

1 What advice would you give to the parents of a new baby concerning footwear and clothes?
2 Explain why a small boy should not wear his older brother's cast-off shoes.

PROJECT WORK

1 Find out how children's feet are measured in a shoe shop, and how the correct size of shoe is determined.
2 Find out if a shop assistant selling children's shoes is given any special training.

9.7 Stand up straight!

This section is concerned with posture. Posture relates to how the body is held. Good posture holds the body with only slight tension in the muscles.

Exercise

This is a short practical exercise.

1 Hold your arm bent slightly at the elbow. Keep it as steady and as comfortable as you can. In this position the muscles are in slight tension.
2 Now hold a weight in the hand, keeping your arm in the bent position. Your arm will soon feel tired.

In this exercise your muscles quickly grew tired from holding the weight. Muscles in your body will also feel tired when your posture is bad.

Consider the weighted toy in Fig. 9.35. When the toy is perfectly balanced, no work or strain is involved in keeping it in position. If the toy is moved, work must be done to hold it still.

Now look at Fig. 9.36(a). The person is perfectly balanced. The weight of the body is transferred to the pelvis and then to the legs. Notice also that the vertebral column is slightly curved. The muscles are working antagonistically to support the body.

Look at Fig. 9.36(b). This person has round shoulders. Notice that the weight of the body is not longer evenly distributed. The back muscles are under tension

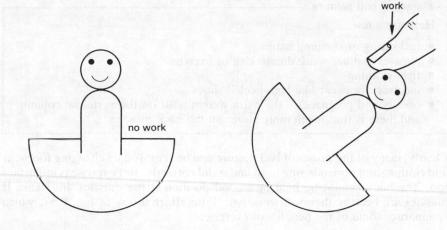

Figure 9.35   Balancing a weighted toy

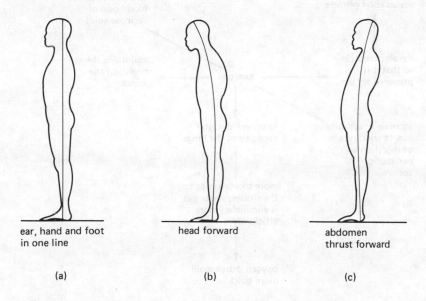

<table>
<tr><td>ear, hand and foot
in one line</td><td>head forward</td><td>abdomen
thrust forward</td></tr>
<tr><td>(a)</td><td>(b)</td><td>(c)</td></tr>
</table>

Figure 9.36   (a) Good posture; (b) round shoulders; (c) hollow back

and will probably ache, and the head is pushed forwards. The chest is cramped, which puts strain on the rib muscles and restricts breathing, and the hepatic portal vein is squashed, giving abdominal discomfort.

Look at Fig. 9.36(c). Notice that the shoulders are held back and the abdomen is pushed forwards to give balance. Someone holding this position is likely to have backache, and also digestive problems because of pressure on the stomach and intestine. Read the boxed note 'Causes of bad posture'.

Causes of bad posture

Here are a few:

- bad sitting or standing habits
- muscles that are weak due to lack of exercise
- tight clothing
- incorrect footwear like high-heeled shoes
- obesity and pregnancy – the extra weight pulls on the vertebral column, and there is thus much more strain on the back muscles.

Clearly, many of the causes of bad posture can be removed by changing footwear and clothing and remembering to sit and stand correctly. But exercise is important too. The body cannot be held in a good position if the muscles are weak. If muscles are healthy they will work with little effort. Look at Fig. 9.37, which summarises some of the benefits of exercise.

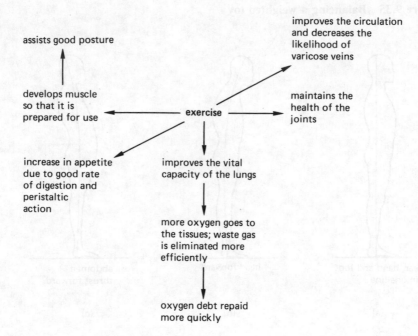

Figure 9.37   The benefits of exercise

Summary

- Good posture holds the body in position with little muscular effort.

- Causes of bad posture:
  lack of exercise, muscles not being used
  sitting or standing badly
  badly fitting shoes
  obesity or pregnancy.

■ Consequences of bad posture:
fatigue
backache
respiratory inefficiency
digestive problems
circulatory inefficiency including varicose veins
lymphatic system inefficiency.

Questions on Section 9.7

1 My job involves sitting at a desk for long hours, but I have had backache since I dug the garden last Sunday. Can you explain why?
2 Look at Fig. 9.38(a). Which muscles are working to lift the box?

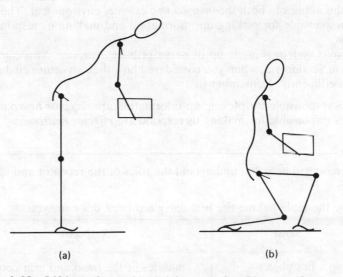

Figure 9.38   **Lifting a heavy box (a) incorrectly, (b) correctly**

Look at Fig. 9.38(b). Which muscles are working to lift the box?
Which is the correct method for lifting the box, (a) or (b)? Explain your answer.

10 Sensing and responding

10.1 Neurones and nerves

In Section 1.5 you learnt that living things are sensitive to their environment. Note that this applies to both the *internal* and *external* environment. The nervous system is responsible for picking up information and making us respond to this information.

The nervous system is made up of nerve cells or *neurones*. You came across nerve cells in Section 1.3, when you considered how their structure enabled them to function efficiently. Remember that:

- neurones responsible for picking up information are *receptor neurones*
- neurones responsible for making us respond are *effector neurones*.

Exercise

This exercise will help you understand the roles of the receptor and effector neurones.

Complete the table below: the first entry has been done for you.

Sense	Response
picking up a hot object	muscles of the hand and arm contract to release the hot object
a fly zooms towards your eye	
you smell some delicious food	
some food get stuck in your throat	
someone calls your name	
an egg is rolling off the table	
you feel cold and have goose pimples	
you feel very thirsty	

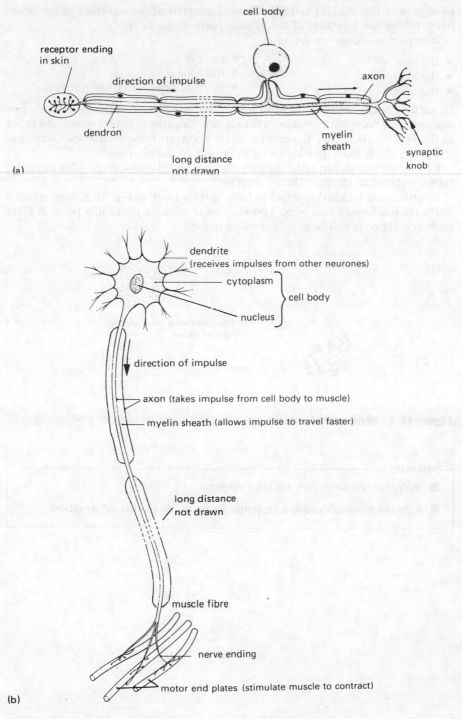

Figure 10.1   (a) A receptor neurone, (b) an effector neurone

Now look at Fig. 10.1(a) and (b). Remind yourself of the structure of the neurones. Notice the functions of the various parts of the cells.

Receptor neurones are found in

- the skin
- the ear
- the nose

- the eye
- the tongue
- internal organs like the intestine

In each of these areas the neurones must be sensitive to different *stimuli*. For example, the receptor neurones in the ear are sensitive to sound waves and those in the nose to chemicals. Receptor neurones convert these stimuli into electrical impulses, which move quickly along the fibres of the neurone.

Effector neurones can make muscles contract and glands secrete – for example, salivary glands to secrete salivary amylase.

Neurones are bundled together to form *nerves*. Look at Fig. 10.2. Notice that a nerve contains many neurones. You may come across a nerve in a piece of meat such as a chop. It will be very soft white tissue.

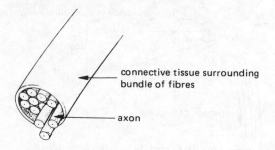

connective tissue surrounding
bundle of fibres

axon

Figure 10.2 Structure of a nerve

┌─ **Summary** ──┐
■ *Receptor neurones* pick up information.

■ *Effector neurones* cause a response to be made to this information.
└──┘

Questions on Section 10.1

Look at Fig. 10.3.

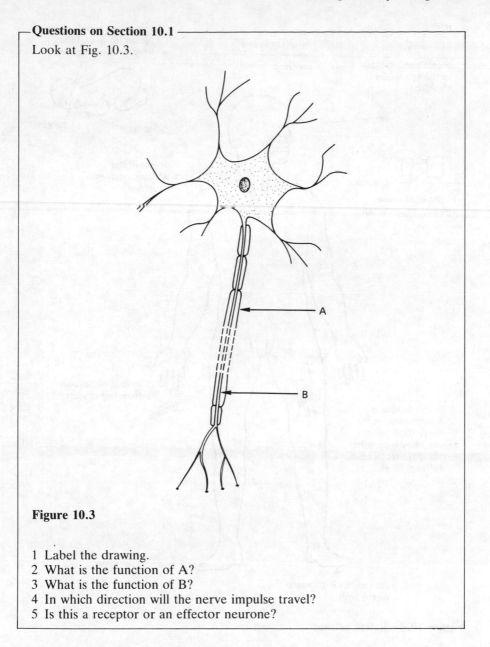

Figure 10.3

1 Label the drawing.
2 What is the function of A?
3 What is the function of B?
4 In which direction will the nerve impulse travel?
5 Is this a receptor or an effector neurone?

10.2 **Responding quickly**

Look at the exercise in Section 10.1. Notice that many of the responses are quick and protect the body from injury. These responses are called *reflex actions*. They are *unlearned* – a newborn baby will withdraw its hand from a hot object. Reflex actions are very quick. Think of your own response to a hot object or pain.

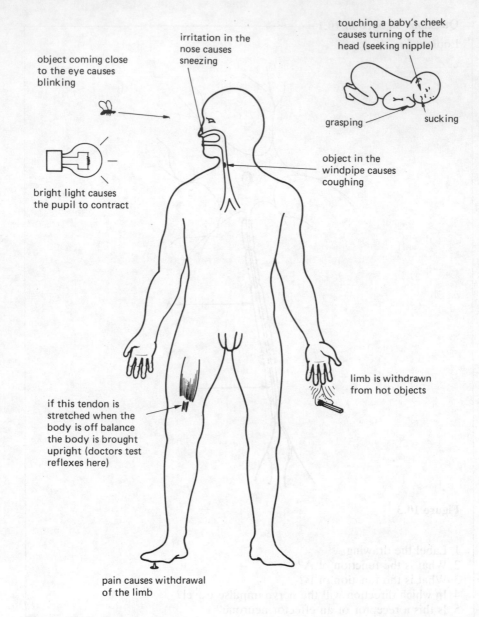

object coming close
to the eye causes
blinking

bright light causes
the pupil to contract

irritation in the
nose causes
sneezing

touching a baby's cheek
causes turning of the
head (seeking nipple)

grasping

sucking

object in the
windpipe causes
coughing

if this tendon is
stretched when the
body is off balance
the body is brought
upright (doctors test
reflexes here)

limb is withdrawn
from hot objects

pain causes withdrawal
of the limb

Figure 10.4 Reflex actions

Let's consider what other reflex actions there are. Look at Fig. 10.4. Notice that babies have several more reflexes than adults have.

For a reflex action to take place there must be at least two components: a *receptor neurone* and an *effector neurone*. Look at Fig. 10.5. Follow the direction of the impulse. Notice that:

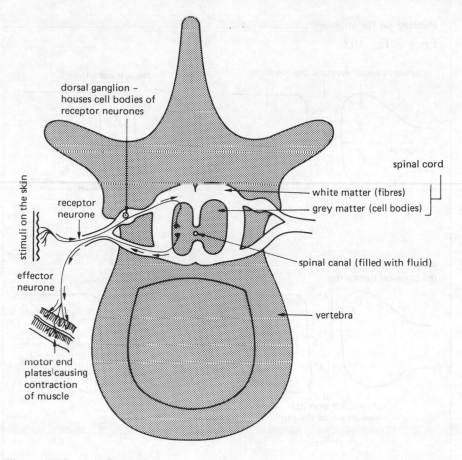

Figure 10.5 A reflex arc

- the receptor and effector neurones *do not* touch
- the cell bodies of the neurones are in the spinal cord.

The brain is not involved in this reflex response, except where the receptors are nearer to the brain than to the spinal cord: the cell bodies of the neurones involved in the reflex action will then be in the brain.

The circuit shown by the arrows in Fig. 10.5 develops as the foetus develops. It is often called a *reflex arc*.

You are probably asking the question, 'if the two neurones do not touch, how does the impulse pass from one neurone to the other?'

To answer this we must look more closely at the junction between the two neurones. Read the boxed note 'Passing on the message'.

Passing on the message

Look at Fig. 10.6.

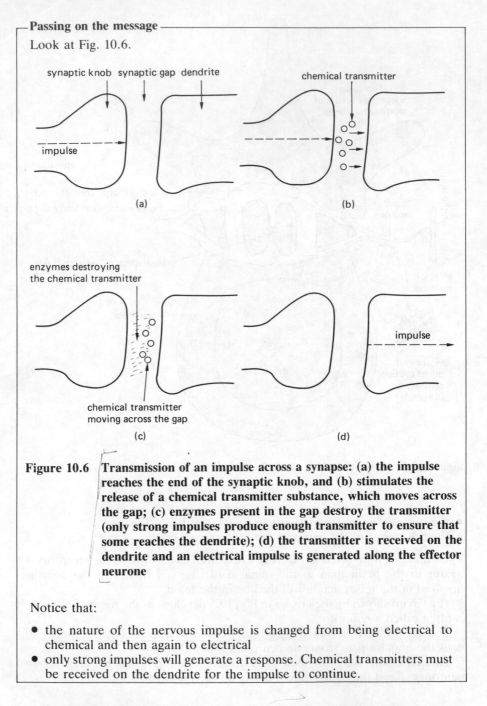

Figure 10.6 **Transmission of an impulse across a synapse: (a) the impulse
reaches the end of the synaptic knob, and (b) stimulates the
release of a chemical transmitter substance, which moves across
the gap; (c) enzymes present in the gap destroy the transmitter
(only strong impulses produce enough transmitter to ensure that
some reaches the dendrite); (d) the transmitter is received on the
dendrite and an electrical impulse is generated along the effector
neurone**

Notice that:

- the nature of the nervous impulse is changed from being electrical to
  chemical and then again to electrical
- only strong impulses will generate a response. Chemical transmitters must
  be received on the dendrite for the impulse to continue.

Look at Fig. 10.7. This is also a reflex arc but it has a third neurone in it, the
connector neurone. The arc works in just the same way as the one in Fig. 10.5.

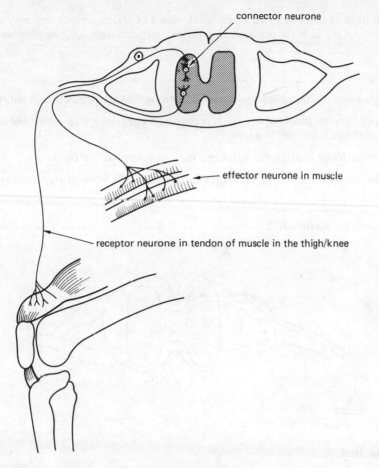

connector neurone

effector neurone in muscle

receptor neurone in tendon of muscle in the thigh/knee

Figure 10.7 A reflex arc with connector neurone

Exercise

Spend a few minutes thinking about the following two situations.

1 You are attending an important meeting. Coffee and biscuits are served. A biscuit crumb gets into your windpipe. You cough to remove it. It remains in your windpipe, however, and you have an urge to cough again. If you cough again it will attract attention, and you cannot leave the meeting.
2 You take a casserole dish out of the oven. The family are all waiting for the meal it contains. It is very, very hot and you have an urge to drop it.

Both these stimuli cause a reflex action, but in both situations you are able to modify your response in some way. Write down how your response may be modified in each case. Are the neurones of the reflex arc the only ones involved in the modified response?

You will have realised that the brain is informed of reflex actions and may cause other responses, like suppressing the reflex, or shouting, or jumping up and down.

Summary

■ *Reflex actions* are unlearned, quick reactions that are protective in nature.

■ The neurones involved are the *receptor neurone*, the *effector neurone* and sometimes a *connector neurone*.

■ The pathway causing the response may not involve the brain.

■ Impulses pass across the synapses in the form of *chemical transmitters*.

Questions on Section 10.2

Look at Fig. 10.8.

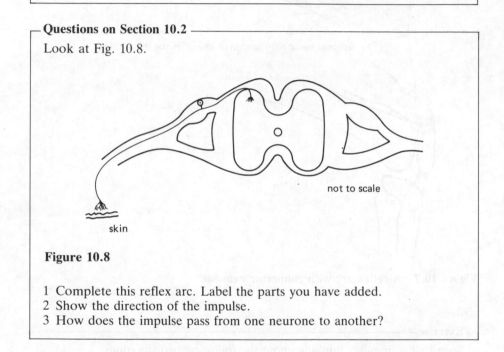

not to scale

skin

Figure 10.8

1 Complete this reflex arc. Label the parts you have added.
2 Show the direction of the impulse.
3 How does the impulse pass from one neurone to another?

10.3 The brain and the body

The brain and the spinal cord form the *central nervous system*. The nerves coming from the central nervous system form the *peripheral nervous system*. Look at Fig. 10.9. Find the brain and spinal cord, and notice that nerves arise from these organs and spread throughout the body. You have come across the brain already (in Sections 7.7 and 8.3), so you should be familiar with its shape and with some of its functions. Let's look at it a little more closely. It is a delicate organ, so it is surrounded by several protective structures. These structures are:

● the skull
● the membranes (called the *meninges*)
● the cerebrospinal fluid.

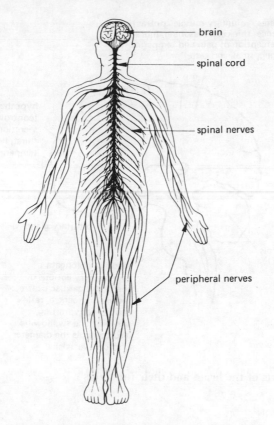

brain

spinal cord

spinal nerves

peripheral nerves

Figure 10.9 The central and peripheral nervous systems

Exercise

This is a practical exercise which will help you to understand the protective role of cerebrospinal fluid.

1 Shake an egg in a screw-top jam jar.
2 Shake an egg in a screw-top jam jar that is full of water.

You will find that the egg in the jar full of water does not break.

The cerebrospinal fluid supports and protects the brain (and spinal cord) and also acts as a cushion or shock absorber between the brain and skull. Sharp blows to the head can of course damage the delicate nervous tissues. Look at Fig. 10.10. Take note of the parts of the brain which you have not come across before: the *cerebral hemispheres*, *cerebellum* and the *medulla oblongata*.

The cerebral hemispheres have many functions. A specific part is responsible for each of these functions. Look at Fig. 10.11. Clearly damage to a particular part of the brain will result in impairment of the function that it controls. Look again at Fig. 10.10. Notice that the surface of the brain is folded. This provides a very large surface area and will allow for many connections between neurones. The range of possible actions and responses is therefore very large indeed.

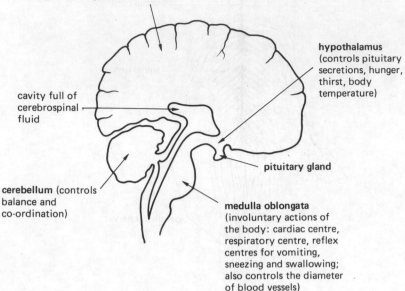

cerebral hemispheres (voluntary muscle contraction, memory, intelligence, thinking, reasoning, moral sense, learning, perception of pain and temperature, touch, sight, hearing, smell and taste)

hypothalamus (controls pituitary secretions, hunger, thirst, body temperature)

cavity full of cerebrospinal fluid

pituitary gland

cerebellum (controls balance and co-ordination)

medulla oblongata (involuntary actions of the body: cardiac centre, respiratory centre, reflex centres for vomiting, sneezing and swallowing; also controls the diameter of blood vessels)

Figure 10.10   Parts of the brain and their functions

┌─ **Summary** ───

■ The functions of the brain are tabulated below:

Part of the brain	Function
cerebral hemispheres	reason, memory, intelligence, thinking, learning, moral sense
cerebellum	balance and co-ordination
medulla oblongata	involuntary actions like breathing, heartbeat, controlling diameter of blood vessels
hypothalamus	controls secretions from the pituitary gland, body temperature, hunger and thirst

■ The brain is protected by cerebrospinal fluid, membranes and the skull bones.

└───

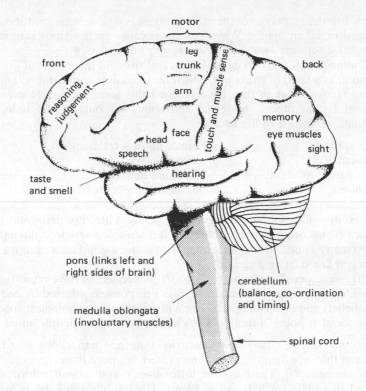

Figure 10.11 Functions of the parts of the cerebral hemispheres

---

Questions on Section 10.3

1 Alcohol slows down the transmission of impulses between neurones. Why are the effects of alcohol on the medulla oblongata particularly serious?
2 A person suffers a blow to the head in a road accident. He discovers that he has lost sensitivity in his hands. Which part of the brain has been damaged in the accident?

10.4 **Stressful lives**

Stress is a part of normal life and is not always bad. For example, getting married, the birth of a baby, buying a new house and getting a new job are normal *life events*. Stressful situations are often self-inflicted!

People perceive and respond to life events in different ways, however. For example, some people thoroughly enjoy moving to a new house and settling in; others find the whole experience a nightmare.

If stress becomes too great it can cause emotional reactions, behavioural changes or physical changes. Look at Fig. 10.12. Of course, the conditions in Fig. 10.12 are not always the result of stress, and stress does not always result in these conditions. But the responses of the body to stress can be serious, and should not be dismissed or taken lightly. Why does stress cause such serious responses in some people? There are two ways of answering this question.

Firstly, some people have to deal with several stressful life events at one time. Each person has a level of stress with which they can deal comfortably, and most people can recognise when they have reached this level. Some life events are more stressful than others, however, and some are outside the individual's control. Look at this list of life events:

death of a spouse	break-up of a relationship
imprisonment	loss of a job
death of a parent or child	personal injury
buying a house	onset of long-term illness

All these events will result in a change in the person's life. Problems may arise if several such events occur at once. Therefore it is sensible to try to avoid having to handle too many at one time – for example, it is not a good idea to move house just when you are starting a new job.

Secondly, some people respond badly to stress because of the circumstances in which the life events occur. For example, one's response is affected by one's own attitudes, beliefs and expectations, by one's physical and psychological condition, and by the social or political climate. Let's have a look at two people under stress:

● Sheila is buying a new house but is worried that the Chancellor's next Budget may mean that her financial commitment may be more than she expected.
● Bob is unemployed. He has very little money and cannot afford a well-balanced diet. His small bedsit is damp. His parents think he is lazy and criticise him for not having a job.

Make sure that you can see the link between the events Sheila and Bob are living through and the factors surrounding their lives.

Look back to Fig. 10.12. Some of the symptoms shown there are related to the release of adrenaline. (You will learn more about this hormone in Section 10.11.) Let's take one of these symptoms – insomnia. Someone who cannot sleep because of stress will find it harder to deal effectively with the life event causing the stress. Clearly, this can lead to anxiety or depression.

The risk of stress-related disease may be reduced if you can recognise the symptoms of stress and if you know how to cope with them. Fig. 10.13 shows some ways of coping, although not all the methods shown will be suitable for everyone. Which methods you adopt depend on your interests and on how much time you have.

Exercise

1 Look at Fig. 10.13. How do you think looking after a pet could help a person cope with stress?
2 (a) How do you cope with stress?
 (b) Could you cope with it more effectively?
 (c) Look at Fig. 10.13 again. How could you improve your own way of coping with stress? What advice would you give to a friend who was finding stress difficult to cope with?

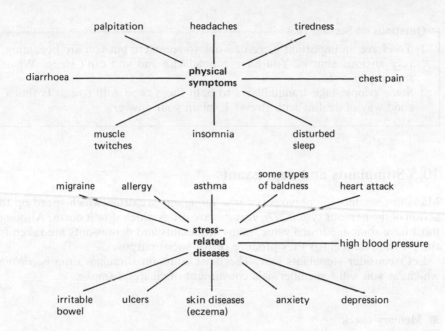

Figure 10.12 Some bodily reactions to stress

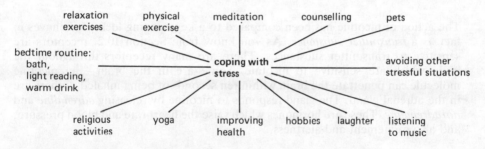

Figure 10.13 Ways of coping with stress

┌─ **Summary** ───────────────────────────────────

■ Normal life events give rise to stress.

■ Individuals have different levels of stress with which they can cope.

■ High levels of stress can cause psychological and physical symptoms. There are diseases related to stress.

■ Responses to stress depend upon the number of life events occurring at any one time and the circumstances in which they occur.

■ Learning to cope with stress may prevent stress-related diseases from developing.
└──

─Questions on Section 10.4 ─────────────────────

1 You have an important piece of work to complete but you are becoming very anxious abut it. You have a headache and you can't sleep. What should you do?
2 Some people take tranquillisers to help them cope with stress. Is this a good way of dealing with stress? Explain your answer.

10.5 Stimulants and depressants

Many drugs affect the nervous system. *Stimulants* are drugs which speed up the action of the nervous system. *Depressants* are drugs which slow it down. Although most have some medicinal value, many stimulants and depressants are taken for the pleasurable feelings they produce or for social purposes.

Let's consider stimulants first. The most common stimulant drug is *nicotine* which, as you will remember, is a constituent of cigarette smoke.

■ **Memory check**
Why is nicotine considered to be a harmful constituent of cigarette smoke?
See Section 7.8.

The action of nicotine has been compared to a key opening locks. It behaves in fact as a *transmitter substance*. As you know from Section 10.2, receptors are sensitive to transmitter substances. There are many receptors throughout the body which are sensitive to nicotine. Some are in the brain – the nicotine molecule can penetrate the brain within ten seconds of being inhaled. Others are in the adrenal gland. The gland responds to nicotine by releasing *adrenaline* and *noradrenaline*. These are hormones which raise the heart rate and blood pressure, and also excitement and alertness.

■ **Memory check**
What is the link between nicotine, adrenaline, high blood pressure and coronary heart disease?
See Section 6.7.

The most common depressant is probably *alcohol*. Like all depressants, it slows down the rate at which nerves transmit information; these depressant effects can be fatal. The action of alcohol on the body is summarised in Fig. 10.14.

You may have wondered why so many people take this drug and why they say it makes them feel relaxed and cheerful. If you look at Fig. 10.14 you will see that the first part of the brain to be affected is the cortex. The cortex normally controls our emotions and allows us to behave in a way that other people find acceptable. If this part of your brain is depressed, you may feel more confident and be able to express your emotions more easily – but of course your abilities are reduced.

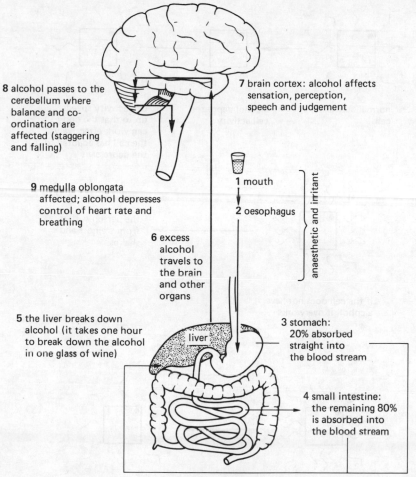

8 alcohol passes to the cerebellum where balance and co-ordination are affected (staggering and falling)

7 brain cortex: alcohol affects sensation, perception, speech and judgement

9 medulla oblongata affected; alcohol depresses control of heart rate and breathing

1 mouth

2 oesophagus

anaesthetic and irritant

6 excess alcohol travels to the brain and other organs

5 the liver breaks down alcohol (it takes one hour to break down the alcohol in one glass of wine)

liver

3 stomach: 20% absorbed straight into the blood stream

4 small intestine: the remaining 80% is absorbed into the blood stream

blood goes to the liver

Figure 10.14   The action of alcohol on the body

Exercise

1 Why is it *not* a good idea to drink alcohol and then drive?
2 Why do people 'feel' that they can control their cars better after they have drunk alcohol?

If a person takes alcohol regularly, the cells of the body become adapted to the drug. Look at Fig. 10.15. The cells become *physiologically dependent* on the alcohol for their normal functioning. Lack of the drug will produce *withdrawal symptoms* such as trembling hands and head.

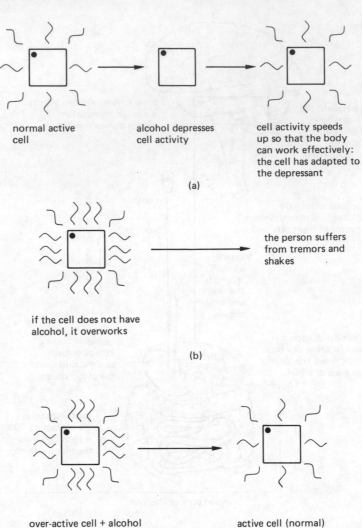

normal active
cell

alcohol depresses
cell activity

cell activity speeds
up so that the body
can work effectively:
the cell has adapted to
the depressant

(a)

if the cell does not have
alcohol, it overworks

the person suffers
from tremors and
shakes

(b)

over-active cell + alcohol

active cell (normal)

(c)

**Figure 10.15   Cell adaptation to alcohol (certain other drugs have similar
effects)**

Exercise

1 What do you think is the difference between psychological and physiological dependence?
2 How would you define 'addiction' to a drug?

Fig. 10.14 shows that the liver breaks down alcohol. But alcohol will damage the liver. In heavy drinkers the liver may become swollen and finally unable to work.

This condition is called liver cirrhosis. Alcohol has also been linked to cancer of the throat; the risk of this cancer developing is increased in people who also smoke.

── Summary ──────────────────────────

■ *Stimulants* speed up the action of the nervous system. *Depressants* slow down the action of the nervous system.

■ *Nicotine* is a stimulant which acts on the receptors in the body in the brain, the adrenal gland and elsewhere.

■ *Alcohol* is a depressant which gives feelings of confidence but reduces abilities severely.

■ Cells of the body can become *adapted* to drugs. Their normal functioning becomes dependent on the presence of the drug.

── Questions on Section 10.5 ───────────────────

1 Look at Fig. 10.16. This is a record of the concentration of alcohol in a man's blood during an evening.

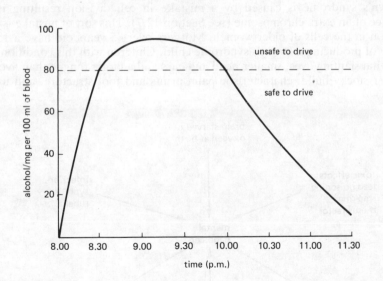

Figure 10.16

(a) According to the graph, what proportion of alcohol in the blood is considered to be safe?
(b) For how long is this man over the legal limit?
(c) At 9.30 p.m. he decides to drive his car home. Do you think he is safe? Explain your reason.
2 It is said that the effects of alcohol are reduced if it is taken with a meal. Why do you think this should be so?

10.6 **Mental handicap**

If the brain is damaged at or before a baby's birth, or if it does not develop properly, the affected person may grow up with physical and/or mental disabilities. In particular, some such people may have difficulties in learning the skills that most people acquire quite easily. These people are often said to be 'mentally handicapped'. Fig. 10.17 shows some of the causes of mental handicap.

The management of mental handicap depends on the severity of the condition; basically, it falls into three areas:

- education, to develop the person's potential
- behaviour modification, where behaviour is disturbed
- (for some people) drug therapy, to control over-active behaviour.

The care provided also depends on the individual's degree of disability and on the home circumstances. Most mentally handicapped children live at home with their parents but as they grow up they may need alternative arrangements. Many can live in the community, with support; some more severely handicapped people may live in residential centres which may be able to provide sheltered employment, such as horticulture or light industry. Look at Fig. 10.18.

There are many types of mental handicap. We are going to look at one type, *Down's syndrome*, in a little more detail.

Down's syndrome is caused by a mistake in cell division resulting in the presence of an extra chromosome (see Section 12.7). This sort of mistake is more common in the cells of older women. Mothers over 45 years old have a 1 in 40 chance of producing a Down's syndrome child. Children with this condition have somewhat slanting eyes, smaller head and ears and a larger tongue than average, a rather stocky build, characteristic palm prints and foot structure, and usually

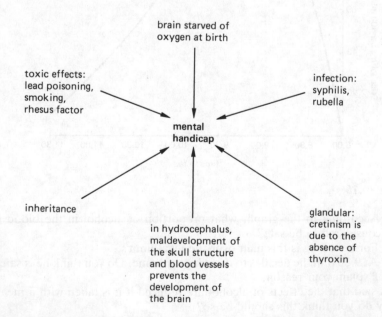

Figure 10.17   Some causes of mental handicap

Figure 10.18 Mentally handicapped adults who enjoy their work: the 'Red Stripe' catering group in North London

marked learning difficulties. With plenty of stimulation, however, a good deal of education is possible.

Summary

■ Brain damage or poor brain development can result in *mental handicap*.

■ The condition causing mental handicap is usually present at birth.

■ Care and treatment depends on the severity of the handicap.

■ *Down's syndrome* is due to the presence of an extra chromosome.

PROJECT WORK

Find out what provisions there are in your district for mentally handicapped people. You may like to consider the following areas:

- small children – education and care facilities
    help for parents
- children of school age – educational and care facilities
    social facilities
    residential centres
- adults – care facilities
    work schemes
    social facilities
    residential centres.

10.7 **Seeing**

Before looking at the eyes in detail, we will consider the various external structures which enable the eyes to function effectively. Look at Fig. 10.19. Notice the mechanism for keeping the eyes clean. Find the tear gland and the eyelid in the drawing. There are also ducts, not shown in the drawing, which drain the tears away. Tears are produced all the time but in larger amounts when the eyes are exposed to certain chemicals or as a result of emotional feelings or pain. The conjunctiva covering the cornea is moistened by the tears and is constantly wiped clean by the eyelids. Tears contain an antimicrobial substance which helps to protect the conjunctiva from infection.

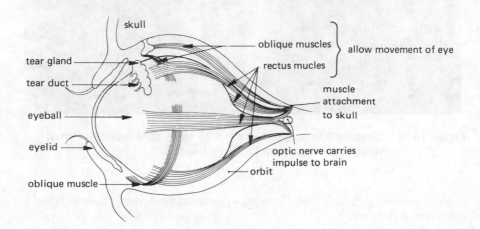

Figure 10.19   The external structures of the eye

Exercise

Look at your eye in a mirror. Draw all you can see. Label the pupil, iris, sclera, eyelash and eyelid.

If you look at the corner of your eye you should see a small hole on your lower eyelid. This is the tear duct which takes away the tears.

Fig. 10.19 also shows the muscles which enable you to move your eyeballs in any direction. Remember that muscles work antagonistically.

■ **Memory check**

What does 'antagonistic' mean in relation to muscles?
See Section 9.5.

Now look at Fig. 10.20, which shows a section of an eye. Find the following:

● the tough protective *sclera* around the eye

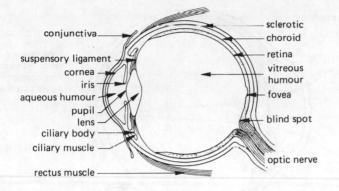

Figure 10.20   Structure of the eye (section)

- the light-sensitive *cornea*, *humours* and *lens* – light must pass through all of these before it reaches the retina
- the *choroid*, which gives rise to the *iris* and *ciliary body*.

Exercise

Make sure you are able to label a diagram of the eye from memory before you go on. Your understanding of how the eye works depends largely on whether you know what all the structures are.

As you know, you can read a book one moment and then look at a cloud in the sky the next. How can the eyes adjust themselves to see objects that are at such different distances from it?

Let's start with a simple exercise.

Exercise

This is a short practical exercise to help you understand how light travels through clear objects.

Put a pencil in a glass of water. Notice that the pencil appears bent. The light rays have been bent or *refracted* as they pass through the water.

Look at Fig. 10.21. Notice that light rays are *refracted* as they pass through the cornea, aqueous humour, lens and vitreous humour. The shape of the lens can change, so changing the amount of refraction.

Let's see how the shape of the lens can be changed. Look at Fig. 10.22(a). This is a surface view of the eye, omitting the iris. Notice that the lens is held by suspensory ligaments. It is the tension of these ligaments that controls the shape of the lens.

- When the ciliary muscle contracts, the diameter of the circle decreases. Look at Fig. 10.22(b). The ligaments slacken and the lens becomes fatter. A fat lens refracts light rays more and enables you to see close objects clearly.

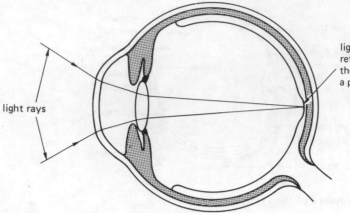

light rays are
refracted so that
they converge to
a point on the retina

light rays

Figure 10.21   The refraction of light rays as they pass through the eye

● When the ciliary muscle relaxes, the diameter of the circle increases. Look at
  Fig. 10.22(c). The ligaments are pulled tight and the lens is pulled thin. A thin
  lens refracts light rays only slightly and enables you to see distant objects
  clearly.

Let us consider the retina, and its sensitivity to bright and dim light. Look back at
Fig. 10.20, and find the retina and the *fovea*. There are two types of light-sensitive
cell in the retina, called *rods* and *cones*. Look at Fig. 10.23. Notice the
characteristics of these two types of cell.

Cone cells are concentrated in the fovea. You are using the cells of your fovea
as you read this page. These give you a clear sharp image in colour.

Look at Fig. 10.24. Notice the distribution of rods and cones in the retina. This
means that if you concentrate an image on the fovea in dim light conditions you
will not be able to see it. Notice too that there are parts of the retina that do not
allow you to see in colour.

The retina contains sensitive receptor neurones that can be easily damaged by
bright light.

┌─ **Exercise** ───┐

Sit in front of a mirror in a bright room. Put your hands over your eyes and
count slowly to 50.

Then remove your hands and look at the pupils of your eyes. You should
see the size of your pupils decrease as they adjust from darkness to
brightness. Try the exercise again if you did not see the change clearly.

└──┘

In this exercise you observed the *reflex action* of your iris which protects the
receptor neurones in your retina from damage by bright light. Look at Fig. 10.25.
Notice that the iris is made up of two sets of muscles. It is the contraction and
relaxation of these muscles which changes the size of the pupils.

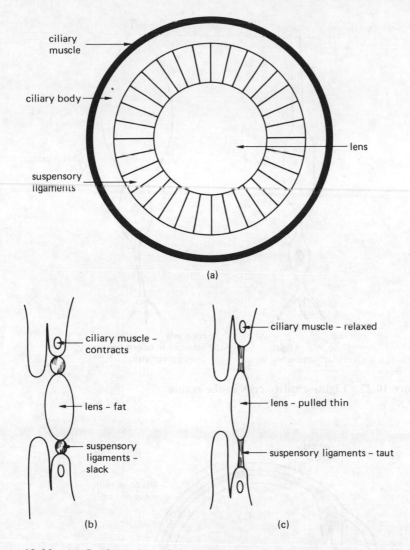

Figure 10.22 **(a) Surface view of the eye; (b) change in shape of the lens (seen in section) to allow vision of near objects; (c) change in shape of the lens to allow vision of distant objects**

Finally, look at Fig. 10.26. Notice that the position of the horse's eyes gives almost all-round vision. This would have been useful to the horse's ancestors, who needed to be aware of the approach of potential predators as they grazed. On the other hand, humans have eyes that face forwards. Their fields of vision overlap, which gives them a good sense of perspective and distance (*stereoscopic vision*). But Fig. 10.26 shows that there is a large area behind the head that humans cannot see. Such a large 'blind area' would be very dangerous for a potential prey such as a horse. But, of course, animals like horses cannot judge distance as well as animals with *binocular vision*.

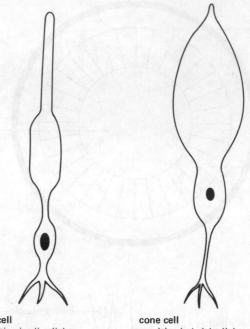

rod cell
sensitive in dim light,
gives black and white vision

cone cell
sensitive in bright light only,
gives colour vision

Figure 10.23   Light-sensitive cells in the retina

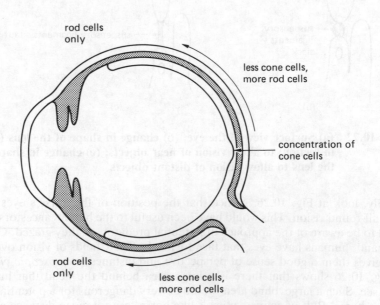

rod cells
only

less cone cells,
more rod cells

concentration of
cone cells

rod cells
only

less cone cells,
more rod cells

Figure 10.24   Distribution of rod and cone cells in the retina

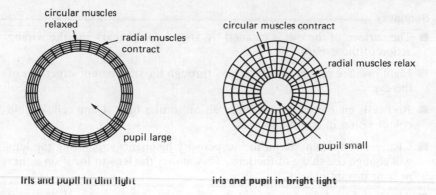

circular muscles relaxed

radial muscles contract

pupil large

circular muscles contract

radial muscles relax

pupil small

Iris and pupil in dim light

iris and pupil in bright light

Figure 10.25   Changes in the size of the pupil in response to bright light

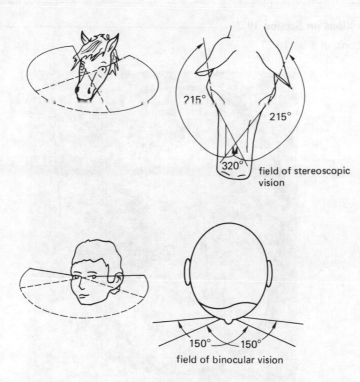

215°

215°

320°

field of stereoscopic vision

150°   150°

field of binocular vision

Figure 10.26   Binocular and stereoscopic vision

── **Summary** ──────────────────────────────────

■ The surface of the eye is cleaned by antimicrobial tears and the wiping action of the eyelids.

■ Light rays are refracted as they pass through the transparent structures of the eye.

■ Rod cells enable black and white vision in dim light. Cone cells enable colour vision in bright light.

■ Changes in the tension on the suspensory ligaments supporting the lens will change the shape of the lens. This allows the lens to focus on either near or distant objects.

■ The muscles of the iris control the diameter of the pupil and therefore how much light can enter the eye.

■ Overlap of the fields of view allows good judgement of distance.

── **Questions on Section 10.7** ──────────────────

1 Look at Fig. 10.27.

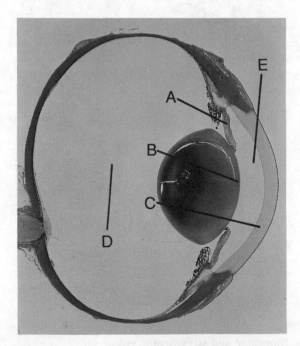

Figure 10.27

(a) What are the structures A, B, C, D and E?
(b) How will the diameter of B alter in going from bright to dim light?
(c) What is the purpose of D?

2 A person is looking directly forwards. A coloured object is moved in the arc shown in Fig. 10.28.

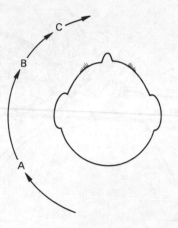

Figure 10.28

(a) Why is the object not visible at A?
(b) Why is the object visible at B, but its colour cannot be seen?
(c) Why can the object be seen in colour at C?

10.8 Hearing

Sound travels through the air as waves. The ear and its hearing mechanism must be able to

- pick up sound waves
- change the waves into nervous impulses.

Let's consider how the ear picks up sound waves.

Look at Fig. 10.29. The ear or *pinna* is designed to pick up the sound waves and direct them down the ear tube. The waves will then cause the *eardrum* to vibrate. The pinna, ear tube and eardrum are all part of the *outer ear*.

The eardrum is responsible for passing on the sound waves. Look at Fig. 10.30. Find the ear tube and the eardrum. Notice that movement of the eardrum will cause the three small bones to knock against each other. The smallest bone, the stirrup, causes movement of the membrane called the *oval window*. The movement of the three bones not only passes on the waves but amplifies them 22 times. These three bones, together with the air-filled cavity bordered by the eardrum and oval window, make up the *middle ear*. Read the boxed note 'Ear popping'.

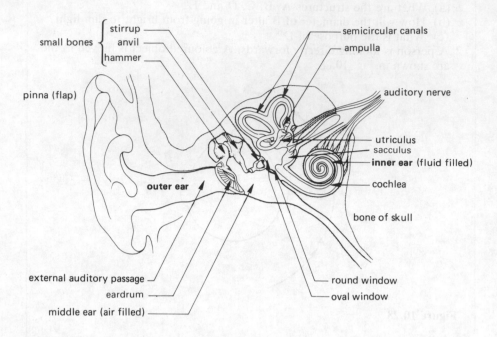

small bones { stirrup, anvil, hammer

pinna (flap)

semicircular canals

ampulla

auditory nerve

utriculus
sacculus
inner ear (fluid filled)

outer ear

cochlea

bone of skull

external auditory passage

eardrum

middle ear (air filled)

round window

oval window

Figure 10.29   Structure of the ear

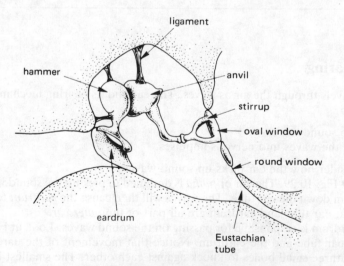

ligament

hammer

anvil

stirrup

oval window

round window

eardrum

Eustachian
tube

Figure 10.30   Structure of the middle ear

Ear popping

It is important that the air pressure on both sides of the eardrum is the same. If you go up in an aeroplane the atmospheric pressure will drop. So the air pushes outwards on the eardrum. The pressure difference is uncomfortable, and may affect hearing because it distorts the eardrum so that it cannot pass on sound waves properly. Swallowing, yawning and chewing can all introduce external air into the *Eustachian tube*, making the 'pop'. This equalises the pressure on both sides of the eardrum. Hearing will then return to normal.

When the aeroplane descends the opposite may happen. The atmospheric pressure will rise until it is higher than the pressure inside the eardrum. The air pushes inwards on the eardrum. Again, introducing air into the Eustachian tube will remove the discomfort and restore normal hearing.

The oval window separates the middle ear from the *inner ear*. Look at Fig. 10.29 again. The inner ear is a fluid-filled cavity in the bones of the skull. In this cavity there is a closed tubular structure, also full of fluid. The snail-like part of the tube is concerned with hearing. This is the *cochlea*. Find the cochlea in Fig. 10.29. Notice that the auditory nerve arises from the cochlea. The cochlea contains the receptor neurones responsible for detecting sound waves. Let's see how they do this.

As the oval window moves it produces ripples in the fluid in the inner ear (like the ripples produced in a washing-up bowl full of water when you tap the outside). The ripples cause movement across the wall of the cochlea, which sets up ripples in the fluid in the cochlea. These ripples disturb sensory hair cells inside the cochlea, and the disturbance generates an impulse in the receptor neurones.

Summary

■ The ear consists of the *outer*, *middle* and *inner ear*.

■ The outer ear collects and directs sound waves into the middle ear.

■ The three small bones of the middle ear amplify and pass the sound waves on to the inner ear.

■ Movement of fluids in the inner ear causes sensory hair cells to move and the receptor neurones are stimulated.

■ The *Eustachian tube* maintains an equal air pressure on both sides of the eardrum.

Look at Fig. 10.31.

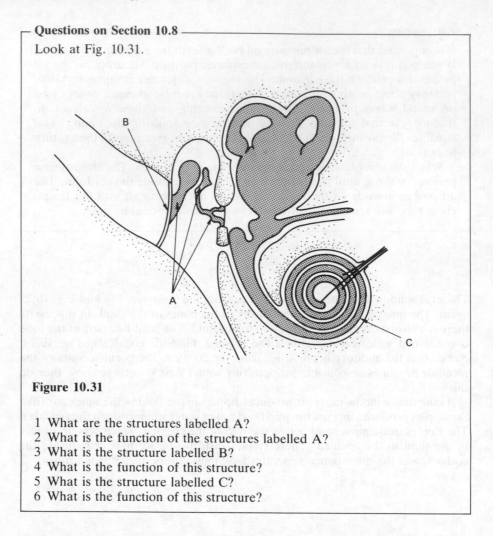

Figure 10.31

1 What are the structures labelled A?
2 What is the function of the structures labelled A?
3 What is the structure labelled B?
4 What is the function of this structure?
5 What is the structure labelled C?
6 What is the function of this structure?

10.9 Knowing where you are: balance

In Section 10.8 you learnt that the cochlea is involved in hearing. If you look back to Fig. 10.29 you will see that the cochlea is only part of the structure in the inner ear. Find the semicircular canals, utriculus and sacculus. These parts of the ear are concerned with balance and position in space – that is, knowing where you are in relation to the ground.

Let's consider the semicircular canals first. Look back at Fig. 10.29. Notice that two of the canals are vertical and at right angles to each other, while the third is in a horizontal plane. Remember that the canals are continuous with the cochlea and that they all contain fluid (*endolymph*). At the end of each semicircular canal there is a swelling called an *ampulla*. Look at Fig. 10.32. Notice that the ampulla contains sensory hair cells embedded in a jelly. The sense cells are responsible for transmitting impulses to the brain regarding the position of the body in space.

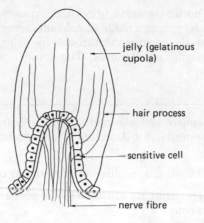

jelly (gelatinous
cupola)

hair process

sensitive cell

nerve fibre

Figure 10.32   Detail of an ampulla

So how does it work? When your head moves, the endolymph in the semi-circular canals also moves. When the endolymph in a canal moves, the cupola in the ampulla is displaced; this pulls on the hairs and stimulates the hair cells. A nerve impulse is then transmitted. Information from each of the ampullae enables you to know where you are in relation to the ground. Look at Fig. 10.33.

Now let's go on to the sacculus and utriculus. When your head is still, the fluid in the semicircular canals is not moving and the ampullae are therefore not stimulated. So the brain is not receiving any information regarding where the body is. The sacculus and utriculus provide this information.

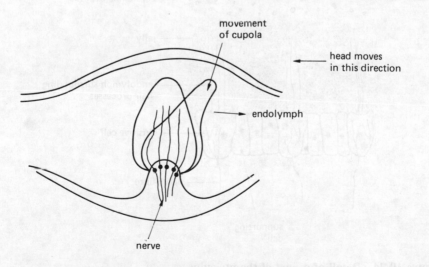

movement
of cupola

head moves
in this direction

endolymph

nerve

Figure 10.33   Action of the ampulla when the head is moved

Look at Fig. 10.34. Notice the nerve fibres and the hairs on the sensitive cells. The *otoliths* are tiny solid particles made of calcium carbonate and they move in response to gravity. As the otoliths fall through the jelly, they distort the hair processes and stimulate the nerve fibres. The result is the transmission of a nervous impulse.

Summary

■ The *semicircular canals*, *utriculus* and *sacculus* are structures in the inner ear which are responsible for detecting where the body is in relation to gravity.

■ The semicircular canals are sensitive to movements of the head.

■ The utriculus and sacculus are sensitive to the position of the head when the body is not moving.

Questions on Section 10.9

If you spin yourself around, it is difficult for you to keep your balance when you stop spinning.

1 What effect does the spinning action have on the structures of the inner ear?
2 Why is it difficult to keep balance when the spinning stops?
3 Which part of the brain will receive the impulses from the inner ear as a result of the spinning?

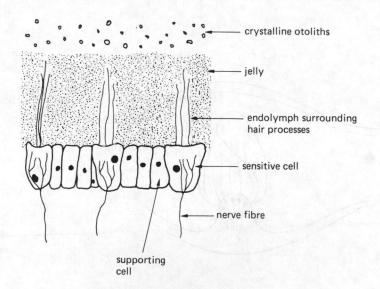

crystalline otoliths

jelly

endolymph surrounding hair processes

sensitive cell

nerve fibre

supporting cell

Figure 10.34   Detail of a part of the utriculus

10.10 **Taste and smell**

In the cavity of your nose there are receptor neurones sensitive to chemicals in the air. Look at Fig. 10.35. Notice that the endings of the receptor neurones are stimulated by chemicals from the air dissolved in the mucus.

■ **Memory check**
Where does the mucus come from?
What is the function of the mucus?
See Section 7.1.

The receptor neurones that enable us to 'taste' are found on the tongue. If you look at your own tongue in a mirror you will notice that it is made up of many small raised portions called *papillae*. Look at Fig. 10.36. Notice that the receptor neurones are on the sides of the papilla. The chemicals in food and drink dissolve in the mucus layer over the tongue. Once they have dissolved they stimulate the sensory endings of the receptor neurones and the impulse is carried to the brain. There are only four types of receptor neurone on the tongue associated with taste, however. These four types are sensitive to chemicals that are respectively bitter, sour, sweet and salty.

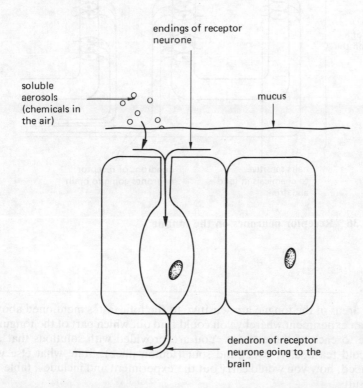

Figure 10.35 Receptor neurones in the nose

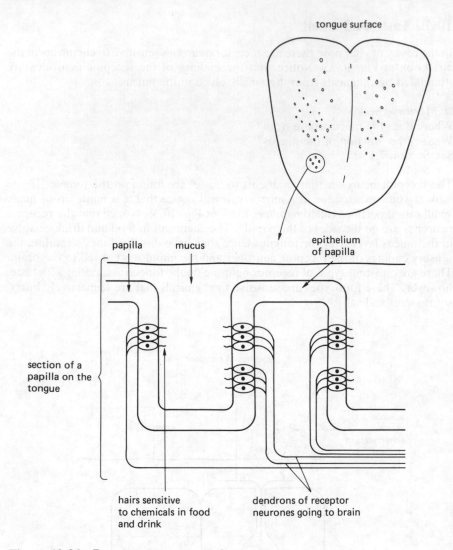

Figure 10.36   Receptor neurones on the tongue

Exercise

Specific areas of the tongue are sensitive to the four tastes mentioned above. Design an experiment whereby you could find out which part of the tongue is sensitive to each of these tastes. You are provided with solutions that are bitter (cold tea), salty, sweet and sour (lemon juice). State what else you would need, how you would carry out the experiment and include a table for your results.

You may be wondering why you are able to detect a whole range of different tastes when you only have four types of taste receptor. Try this next exercise to help you understand why this is so.

Exercise

This is a short practical exercise. You will need a colleague to help you, and some cubes of food all cut to the same size – try apple, raw potato and pear.

Close your eyes and pinch your nose so that you cannot see or smell. (You could use a blindfold and a nose clip.) Get your colleague to put a cube of food on to your tongue. See if you can tell what it is.

Repeat the experiment but this time you do not have to pinch your nose.

You will have found that if you cannot smell, your sense of taste is spoilt. Smell enables us to experience a wide range of 'tastes'. In other words, the two systems work together.

Summary

■ Our senses of *smell* and *taste* depend on chemicals dissolving in mucus and then stimulating the endings of receptor neurones.

■ There are four types of receptor neurone responsible for taste. They are found in specific areas on the tongue.

■ Smell and taste work together to give us sensitivity to a wide range of tastes.

Questions on Section 10.10

1 Explain why a person with a heavy cold complains that 'everything tastes the same'.
2 A chemical must be soluble if we are to smell or taste it. Explain why this is so.

10.11 **Responding slowly**

The *endocrine system* consists of *endocrine glands* in different parts of the body. But not all glands are endocrine glands. Read the boxed note 'Two sorts of gland'.

Hormones regulate and co-ordinate various functions in the body. Let's examine what hormones do. Look at Fig. 10.39. Notice that a hormone is produced by one organ but has its effect on one or more other organs, called its *target organs*. For example, insulin is produced in the pancreas, but its target organ is the liver.

Two sorts of gland

Look at Fig. 10.37. The salivary gland has a structure similar to this. Its cells produce a secretion which then passes out of the gland via a duct.

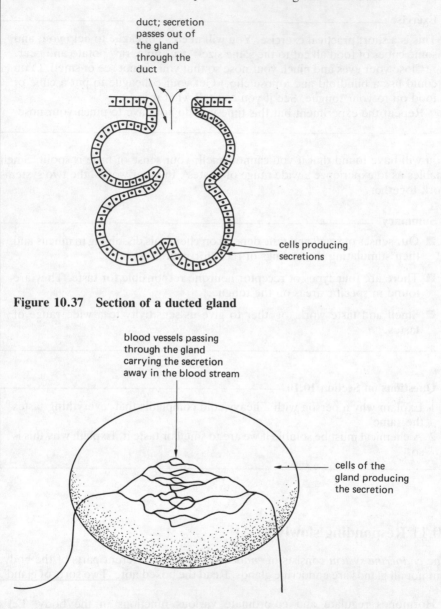

duct; secretion passes out of the gland through the duct

cells producing secretions

Figure 10.37   Section of a ducted gland

blood vessels passing through the gland carrying the secretion away in the blood stream

cells of the gland producing the secretion

Figure 10.38   A ductless (endocrine) gland

Now look at Fig. 10.38. This is an endocrine gland: its secretion is passed into the blood stream. Endocrine glands produce *hormones*, which are sometimes called *chemical messengers*.

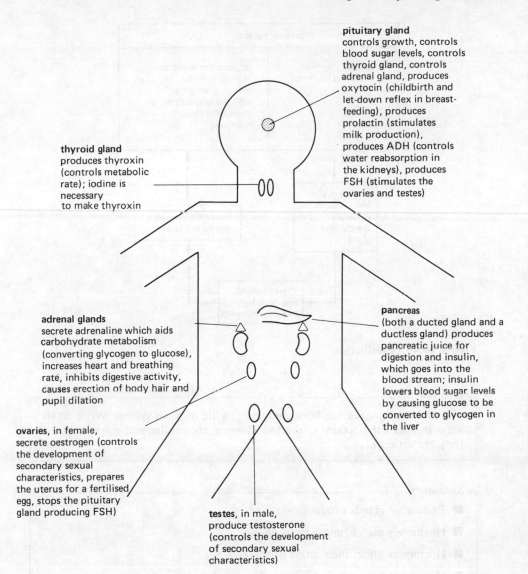

pituitary gland
controls growth, controls blood sugar levels, controls thyroid gland, controls adrenal gland, produces oxytocin (childbirth and let-down reflex in breast-feeding), produces prolactin (stimulates milk production), produces ADH (controls water reabsorption in the kidneys), produces FSH (stimulates the ovaries and testes)

thyroid gland
produces thyroxin (controls metabolic rate); iodine is necessary to make thyroxin

adrenal glands
secrete adrenaline which aids carbohydrate metabolism (converting glycogen to glucose), increases heart and breathing rate, inhibits digestive activity, causes erection of body hair and pupil dilation

pancreas
(both a ducted gland and a ductless gland) produces pancreatic juice for digestion and insulin, which goes into the blood stream; insulin lowers blood sugar levels by causing glucose to be converted to glycogen in the liver

ovaries, in female, secrete oestrogen (controls the development of secondary sexual characteristics, prepares the uterus for a fertilised egg, stops the pituitary gland producing FSH)

testes, in male, produce testosterone (controls the development of secondary sexual characteristics)

Figure 10.39 The action of endocrine glands

The endocrine gland needs to be aware of the effects of the hormone its produces. Look at Fig. 10.40. This shows how the pituitary gland is controlled by the levels of thyroxin in the blood. This kind of mechanism is called *feedback*.

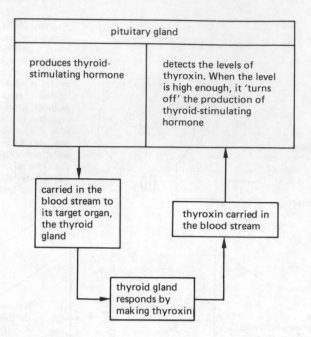

Figure 10.40   Feedback

Exercise

The endocrine system has been compared to the nervous system. Write down ways in which the actions of the two systems are similar and ways in which they are different.

Summary

- *Endocrine* glands produce *hormones*.

- Hormones are chemical messengers.

- Hormones affect their *target organs*.

- Hormones get to their target organs via the blood stream.

- The endocrine glands control their own secretions as a result of *feedback*.

Questions on Section 10.11

1 Adrenaline is sometimes called the 'fight or flight' hormone. Why?
2 Explain how the action of an endocrine gland like the pituitary gland is different from that of a gland like the salivary gland.
---

11 Reproduction

11.1 Sex isn't easy!

You saw in Section 1.5 that one of the characteristics of living things is the ability to reproduce. Plants and animals can do this in all sorts of ways – splitting in two, bits breaking off and developing into new organisms, or producing special cells which must then fuse together to form the beginning of a new individual. Human beings reproduce by the third of these methods, which is referred to as *sexual reproduction*. The special cells are called *gametes* or *sex cells*.

Look at Fig. 11.1. Notice that the terms used to describe the gametes from the male and female and the fusion of the two cells. Human beings are mammals, and in mammals fertilisation and development is *inside* the female.

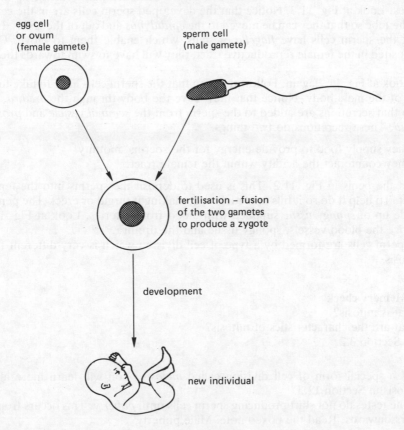

egg cell
or ovum
(female gamete)

sperm cell
(male gamete)

fertilisation – fusion
of the two gametes
to produce a zygote

development

new individual

Figure 11.1 Sexual reproduction (diagrammatic – not to scale)

So humans must be able

- to make gametes
- to get the gametes together so that fertilisation can take place
- to provide a suitable place for development.

In the next few sections we will look at how humans carry out these functions.

11.2 **The male system**

Let's start with where the male gametes are made. Look at Fig. 11.2. Notice that the sperm cells are made in the *testes*, which are held away from the body. This means that the testes are slightly cooler than the rest of the body. Healthy sperm cells need this lower temperature.

Exercise

Why might tight trousers or underpants be a reason for infertility in a male?

The sperm cells are made within fine tubes which are tightly coiled up in the testes. Look at Fig. 11.3. Notice that the developed sperm cells are in the centre of the tube so that they can be moved to the *epididymis* and out of the testes. Note that the sperm cells have *flagella* or 'tails' which enable them to move. Once deposited in the female reproductive tract, they will have to swim towards the egg cell.

Look at Fig. 11.2 again. Follow the path that the sperm cells have to take to get out of the male body. Notice that they leave the body through the *urethra*, and also that secretions are added to the sperm from the *seminal vesicle* and *prostate gland*. These secretions do two things:

- they supply food to provide energy for the sperms' motility
- they counteract the acidity within the female tract.

Find the penis in Fig. 11.2. This is used to deposit the sperms into the female tract. To help it do so it fills up with blood, making it turgid or erect. The penis is made up of *spongy tissue* supplied with blood from arteries. Look at Fig. 11.4. Notice the blood vessels, spongy tissue and the urethra.

Sperm cells are formed by a type of cell division which is very different from mitosis.

■ **Memory check**
What is mitosis?
What are the characteristics of mitosis?
See Section 3.2.

It is a special form of cell division called *meiosis*. You will learn more about meiosis in Section 12.3.

The testes do not start producing sperm cells until *puberty*. This occurs from 13 years onwards. Read the boxed note 'Male puberty'.

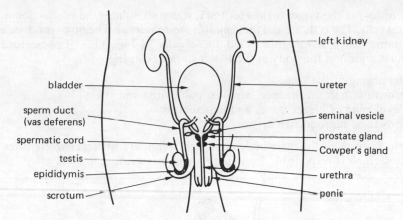

male reproductive organs (position in body)

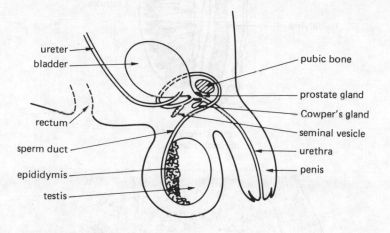

male reproductive organs (vertical section)

Figure 11.2   The male reproductive system

Male puberty

Puberty begins when the pituitary gland starts producing a hormone called FSH, which is short for *follicle stimulating hormone*.

■ **Memory check**
What is a hormone?
How do hormones get to their target organs?
Where is the pituitary gland?
See Section 10.11.

The testes are the target organs for FSH, which stimulates the production of sperm cells. The cells around the *seminiferous tubules* also begin to produce a hormone called *testosterone*. Find these cells in Fig. 11.3. Testosterone circulates around the body in the blood stream, causing

- deepening of the voice
- growth of hair on the face, armpits, pubic area and chest
- developing of bigger bones and musculature
- growth of reproductive organs.

These features of the male body are called *secondary sexual characteristics*. Testosterone is responsible for initiating and maintaining them.

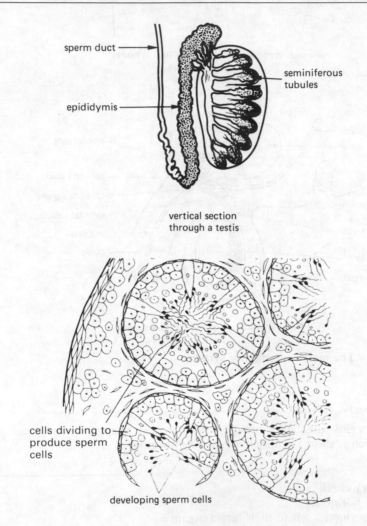

Figure 11.3   Structure of the testis

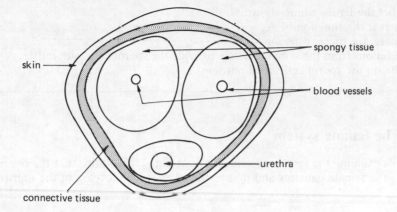

skin

spongy tissue

blood vessels

urethra

connective tissue

Figure 11.4   Section of a penis

Summary

■ *Sperm cells* are male gametes.

■ Sperm cells are made in the *testes*.

■ Secretions from the *seminal vesicle* and *prostate gland* are added to the sperm.

■ The *penis* is made of spongy tissue which fills up with blood. The resulting erect penis can deposit the sperm cells into the female tract.

■ *Puberty* starts when the pituitary gland begins to produce FSH.

■ *Testosterone* causes the development and maintenance of *secondary sexual characteristics*.

Questions on Section 11.2

Look at Fig. 11.5.

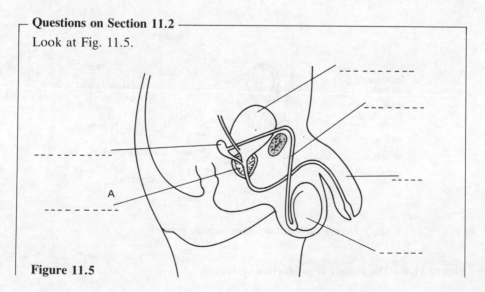

A

Figure 11.5

1 Label the figure where shown.
2 What is the function of A?
3 Where are sperm cells made?
4 Secretions from the seminal vesicle provide the sperm cells with nutrients. Why do the sperm cells need nutrients?

11.3 **The female system**

Now let's examine the female system. Look at Fig. 11.6. Notice that the ovaries produce the female gametes and that fertilised eggs will develop in the uterus.

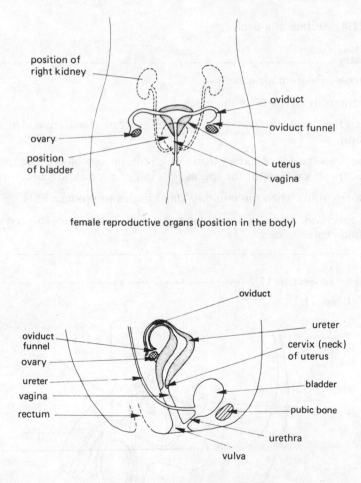

female reproductive organs (position in the body)

female reproductive organs (vertical section)

Figure 11.6   The female reproductive system

Ovaries produce egg cells in cycles which usually last about 28 days. We are going to follow one of these cycles. Look at Fig. 11.7. Notice that the cycle begins with the production of FSH. The ovary is the target organ for FSH and it responds in several ways.

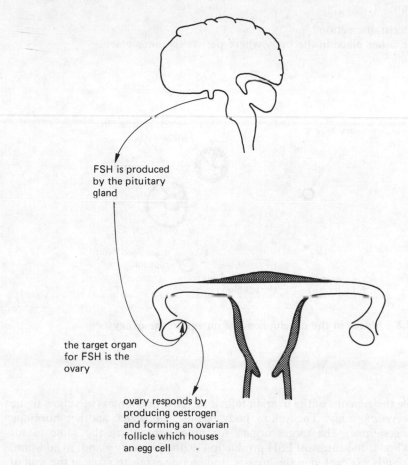

FSH is produced by the pituitary gland

the target organ for FSH is the ovary

ovary responds by producing oestrogen and forming an ovarian follicle which houses an egg cell

Figure 11.7 The first stage of the menstrual cycle

- An egg cell begins to develop. Look at Fig. 11.8. Notice that over the course of fourteen days the egg develops in a blister or bulge on the side of the ovary, called an *ovarian follicle*.
- It produces the hormone *oestrogen*. The levels of this hormone increase over the first fourteen days of the cycle. It turns off FSH production by the pituitary gland, causes ducts to develop in the breasts and also causes the walls of the uterus to thicken. These effects serve to prepare the body for a fertilised egg.

On the fourteenth day, oestrogen levels reach a peak and the ovarian follicle breaks open. The egg cell is released into the oviduct – this is called *ovulation*. The egg cell is then carried along the oviduct by movement of cilia and peristaltic action.

■ **Memory check**

What are cilia?

Name one place in the body where you can find cilia.

What is their function there?

See Section 7.1.

What is peristaltic action?

Name one other place in the body where peristalsis takes place.

See Section 5.6.

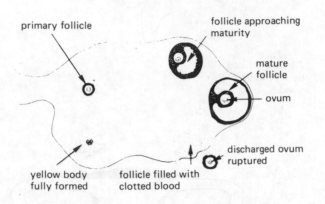

Figure 11.8   Stages in the development of an egg in the ovary

Meanwhile the remains of the ovarian follicle develop into a mass of yellow tissue called the *yellow body*. The yellow body begins to produce another hormone called *progesterone*. The target organs for progesterone are the same as for oestrogen and it too turns off FSH production by the pituitary gland; in addition, it causes glands to develop in the breasts, and blood vessels to grow in the wall of the uterus.

If the egg is not fertilised the yellow body will begin to break down. This will reduce progesterone levels and therefore

• the pituitary gland is no longer prevented from producing FSH
• breast tissue will be reabsorbed
• the lining of the uterus is no longer maintained and so it falls away. This is the *menstrual flow*.

Look at Fig. 11.9. Notice that as progesterone levels increase oestrogen levels fall. Read over the text in this section again, and this time refer to Fig. 11.9 as you do so.

You may have guessed that FSH production in the female begins at puberty. Read the boxed note 'Female puberty'.

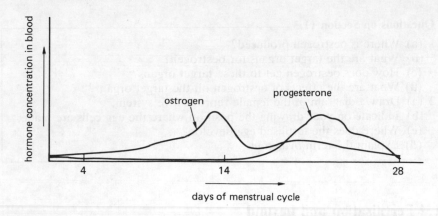

Figure 11.9 Hormone levels during the menstrual cycle

Female puberty

Oestrogen is responsible not only for the changes already mentioned, but also for:

- growth of hair in the armpits and pubic area
- widening of the hips
- deposition of fat under the skin
- development of breast tissue.

These are the *secondary sexual characteristics* of the female, and start to appear after the beginning of puberty.

Summary

■ Female gametes are called *egg cells* or *ova*.

■ Egg cells are made in the *ovaries*.

■ FSH stimulates the production of egg cells.

■ The ovary produces *oestrogen* which causes the lining of the uterus to thicken and breast tissue to develop, and turns off the FSH production from the pituitary gland.

■ Release of an egg cell from the ovary is called *ovulation*.

■ The ruptured ovarian follicle develops into a *yellow body* which produces progesterone.

■ *Progesterone* continues the work of oestrogen in preparing the body for a fertilised egg.

■ If the egg is not fertilised, the yellow body breaks down, progesterone levels fall and the prepared uterine lining falls away. The *menstrual cycle* then starts once more.

11.4 **Fertilisation and beyond**

You now know where the sex cells or gametes are made. In this section we are
going to consider how the cells get together.

In Section 11.2 you learnt that the penis can become erect so that it can deposit
sperm cells in the female system. Look at Fig. 11.10. Notice that the sperm cells
are released from the penis at the cervix. They must then swim through the uterus
and along the egg tubes towards the eggs. Look at Fig. 11.10 again. Notice the
position where the egg cells are fertilised. The sperm cells have to travel a
considerable distance before they reach the eggs.

Once the sperm cells reach the egg cell they try to penetrate the cell surface
membrane. Look at Fig. 11.11. Finally one sperm cell does penetrate the
membrane and its nucleus combines with that of the egg cell. Changes then take
place in the membrane which prevent any more sperm cells from penetrating.

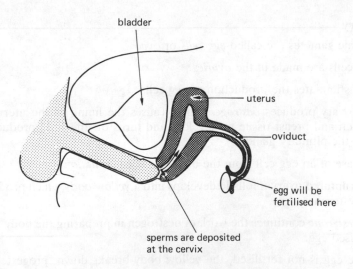

**Figure 11.10   Copulation resulting in the deposition of sperm cells in the female
tract**

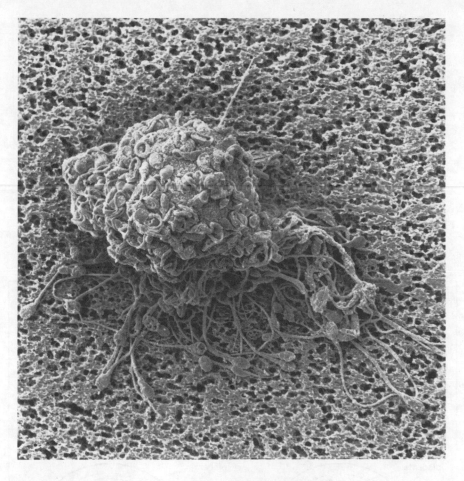

Figure 11.11   Sperm cells clustering around an egg cell

The fertilised egg quickly produces a hormone which maintains the yellow body. Consequently progesterone levels are maintained. Look at Fig. 11.12. Compare this with Fig. 11.9. Notice that if an egg is fertilised, progesterone levels do not fall.

The fertilised egg is now moved slowly towards the uterus. As it is being moved, the single cell is dividing by mitosis. (Read the boxed note 'Multiple births'.)

■ **Memory check**
What is mitosis?
See Section 3.2.

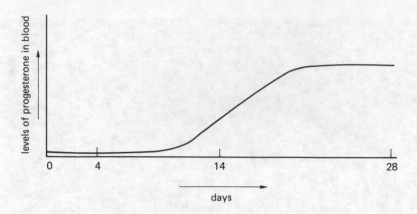

Figure 11.12   Levels of progesterone if an egg is fertilised

Multiple births

Sometimes a woman produces more than one egg at a time. If they are all fertilised, each zygote may develop its own placenta and grow as a single foetus. The result is *fraternal twins* (or triplets or quadruplets if three or four eggs are fertilised). The babies will not be identical and may be of different sexes.

Identical twins are formed when *one* egg is fertilised to produce a zygote and the zygote then divides to give two cells, each of which behaves as if it is a zygote. Two foetuses will develop. Identical twins are always of the same sex. Look at Fig. 11.13. Notice that identical twins may either share one placenta, or have two very close together.

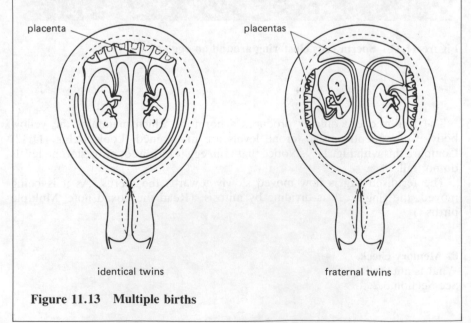

Figure 11.13   Multiple births

Look at Fig. 11.14. Notice that when the ball of cells arrives at the uterus it sinks into the lining. This is called *implantation*. Notice how the *foetus* (the immature baby in the uterus) develops from the ball of cells; so too do the fluid that surrounds it throughout the pregnancy (the *amniotic fluid*) and the *umbilical cord* through which it receives nutrients and gets rid of waste products.

In Fig. 11.14(c) you can see small processes growing into the wall of the uterus. These processes develop into the *placenta*. Eventually many tiny *microvilli* penetrate the uterus wall, as shown in Fig. 11.14(d). They provide an enormous area over which nutrients reach the foetus, and the foetus gets rid of waste products.

Exercise

1 Where does the foetus get its nutrients from before the placenta has developed?
2 Why does the foetus need a placenta as it gets bigger?

Let's look at the placenta in a little more detail. Look at Fig. 11.15. Notice which substances are exchanged across the wall of the placenta. You should also note that the mother's blood and the baby's blood do not mix – they are kept quite separate.

The placenta has several important protective functions:

- it acts as a barrier, preventing microbes passing from mother to baby and causing disease
- it prevents the mother's blood entering the foetus, which could be serious if the blood groups are different (see Section 6.5)
- it prevents damage to the foetus from the mother's higher blood pressure.

Exercise

There are some substances against which the placenta is an inefficient barrier. Spend a few minutes listing some harmful substances that can pass from the mother to the baby.

Summary

■ *Fertilisation* occurs when an egg and a sperm cell fuse to produce a *zygote*.

■ The zygote divides by mitosis.

■ The resulting ball of cells implants into the wall of the uterus.

■ A *placenta* develops which protects the foetus and over which nutrients and waste can be exchanged.

■ *Multiple births* occur when more than one egg is fertilised at the same time, or when a fertilised egg divides and each cell of the cells thus formed develops into a foetus.

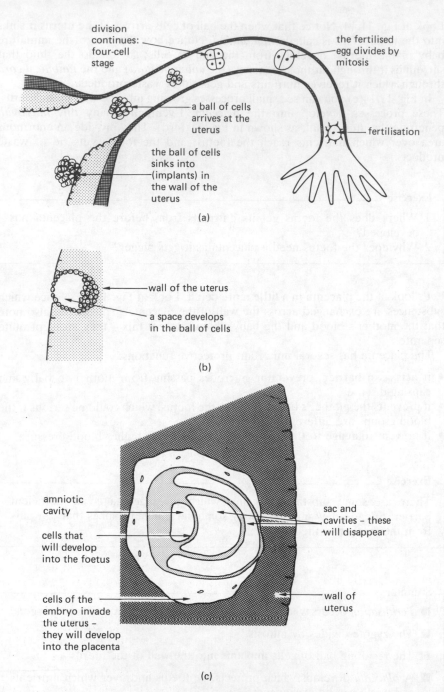

division
continues:
four-cell
stage

the fertilised
egg divides by
mitosis

a ball of cells
arrives at the
uterus

fertilisation

the ball of cells
sinks into
(implants) in
the wall of the
uterus

(a)

wall of the uterus

a space develops
in the ball of cells

(b)

amniotic
cavity

sac and
cavities – these
will disappear

cells that
will develop
into the foetus

cells of the
embryo invade
the uterus –
they will develop
into the placenta

wall of
uterus

(c)

Figure 11.14   Stages in the development of the fertilised egg

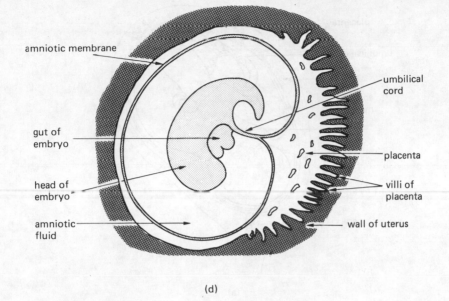

amniotic membrane

umbilical
cord

gut of
embryo

placenta

head of
embryo

villi of
placenta

amniotic
fluid

wall of uterus

(d)

Figure 11.14   (*continued*)

Questions on Section 11.4

1 What is a placenta?
2 What are the functions of the placenta?
3 Why should pregnant women neither smoke nor drink alcohol?

11.5 **Mother and baby**

A pregnant woman is encouraged to attend a clinic for pre-natal tests. These tests
will check that the mother is well and that the baby is healthy and growing. Tests
are carried out monthly until the thirtieth week of the pregnancy, fortnightly up to
36 weeks and then weekly until the baby is born, usually at about 40 weeks.

Look at Fig. 11.16. This shows the tests given to all pregnant women. Notice
that it is irregularities in a steady pattern that can indicate that something is going
wrong. Some irregularities indicate that the woman may be developing a
condition called *toxaemia*, in which the body retains too much water. Find out
from Fig. 11.16 which tests may indicate toxaemia. The exact causes of the
condition are unknown, but if it is not diagnosed and treated it can be fatal.

Sometimes babies are born with a disease or abnormality. It is possible to find
out if a baby has certain diseases or abnormalities by conducting further antenatal
tests. Usually the tests are only carried out if it is believed that there is a high risk
that the baby is not developing normally.

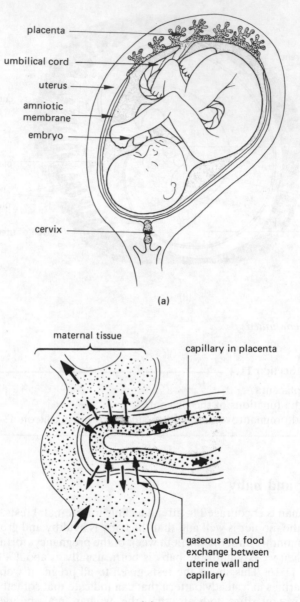

placenta

umbilical cord

uterus

amniotic membrane

embryo

cervix

(a)

maternal tissue

capillary in placenta

gaseous and food exchange between uterine wall and capillary

(b)

Figure 11.15   (a) The placenta and its relationship to the foetus; (b) detail of a capillary in a villus of the placenta

One of the most frequently used tests is *ultrasound scanning*. High-pitched sound waves are directed at the foetus. The reflected waves are changed into a recognisable shape which can be viewed on a screen. Usually this is used if:

- the age of the foetus is in doubt, so that the possible date of birth can be determined

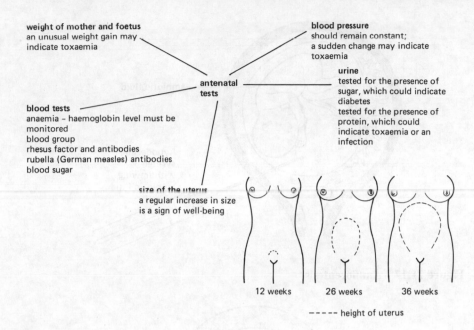

weight of mother and foetus
an unusual weight gain may
indicate toxaemia

blood pressure
should remain constant;
a sudden change may indicate
toxaemia

urine
tested for the presence of
sugar, which could indicate
diabetes
tested for the presence of
protein, which could
indicate toxaemia or an
infection

antenatal
tests

blood tests
anaemia – haemoglobin level must be
monitored
blood group
rhesus factor and antibodies
rubella (German measles) antibodies
blood sugar

size of the uterus
a regular increase in size
is a sign of well-being

12 weeks   26 weeks   36 weeks

- - - - - height of uterus

Figure 11.16   Antenatal tests

- there has been bleeding during pregnancy, which means that the placenta may be lying over the cervix
- twins or multiple births are suspected
- breech presentation (the child lying with its head upwards instead of downwards) is suspected.

The test will also confirm that growth is steady and progressing well.

Amniocentesis is another test given to pregnant women, but it is only carried out on women whose babies are at high risk. Some mothers lose their babies (miscarry) as a result of the test. Look at Fig. 11.17. Notice that amniotic fluid is withdrawn. The fluid contains foetal cells. Examination of the cells, particularly their chromosomes, can reveal the following:

- Down's syndrome, which is most common in the children of women over 40 and in children whose mothers already have a Down's syndrome child
- other inherited defects
- the child's sex, which is important in families in which there are inherited diseases which affect boys only (haemophilia is an example)
- foetal proteins, which indicate a spinal cord condition such as spina bifida.

Biopsy of chorionic villi involves removing a little tissue from the villi of the placenta. Examination of these cells gives the same information as amniocentesis but the test can be performed several weeks earlier in the pregnancy. It also carries the risk of miscarriage, however. Look at Fig. 11.18.

If either amniocentesis or the biopsy procedure indicates that the baby will be born with a serious disability, the mother may choose to have her pregnancy ended.

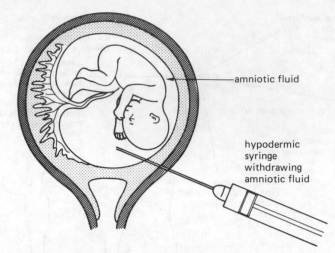

amniotic fluid

hypodermic
syringe
withdrawing
amniotic fluid

Figure 11.17   Amniocentesis

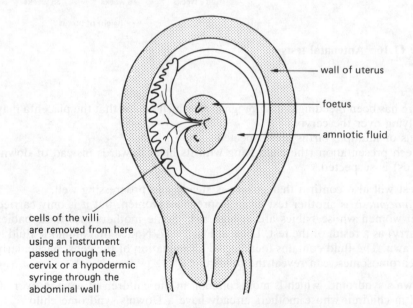

wall of uterus

foetus

amniotic fluid

cells of the villi
are removed from here
using an instrument
passed through the
cervix or a hypodermic
syringe through the
abdominal wall

Figure 11.18   Biopsy of chorionic villi

In most cases a baby develops perfectly normally. It grows rapidly; during the pregnancy the uterus expands to 24 times its original size and reaches up to the base of the breast bone. The foetus is supplied with oxygen and nutrients by blood coming from the placenta. It also gets rid of wastes (mainly urea and carbon dioxide) through the placenta. So its lungs, gut and liver do not function as they will do when the baby is born. The blood supply to these organs is quite small at this stage.

Look at Fig. 11.19, which shows the foetal circulation. Find the heart (not all the vessels are drawn). The pulmonary artery and aorta are shown by dotted lines in the heart. Make sure you know what the various areas of the heart are. Refer back to Section 6.3 if necessary. There are several things to note in Fig. 11.19.

- There is a placental circulation (the umbilical arteries, umbilical vein and the placenta itself).
- The umbilical vein takes blood to the foetus' hepatic portal vein. The *ductus venosus* takes most of the blood from the hepatic portal vein to the vena cava. Only a little of the blood goes to the liver.

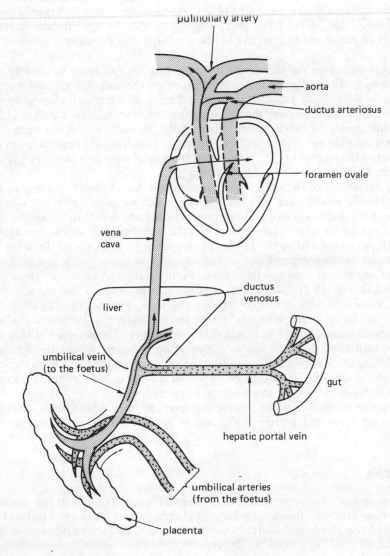

Figure 11.19 The foetal circulation before birth

- Blood in the vena cava passes through the *foramen ovale* in the septum of the heart, into the left atrium. Most blood bypasses the lungs, which are not working as oxygenators yet.
- The *ductus arteriosus* takes blood from the pulmonary artery to the aorta.

Birth

At the end of the pregnancy the foetus is usually lying with its head down and its arms and legs folded close to its body. The next stage in its life is birth.

The first stage in the process is the release by the pituitary gland of a hormone called *oxytocin*, which causes the muscles of the uterus to contract. The first contractions may dislodge the mucus plug that sealed the cervix during pregnancy and protected the uterus from micro-organisms. So the appearance of small quantities of blood and mucus may be the first sign that the baby is about to be born.

Oxytocin will cause the muscles of the uterus to contract more frequently and more strongly. These contractions put pressure on the amniotic sac and force it downwards. Under this pressure the cervix begins to dilate and eventually the amniotic sac breaks, releasing the amniotic fluid or 'waters'. Look at Fig. 11.20. Notice how much the cervix has to dilate in the second stage of the birth.

When the cervix has dilated to about 10 cm, the contractions begin to push the foetus out of the mother's body. Notice that the foetus must turn at right angles to get through the mother's pelvis and vagina.

The contractions of the uterus now squeeze the foetal blood standing in the placental blood vessels back into the foetus – which we should now call a baby. The placental circulation stops. From now on the baby needs to obtain its own oxygen from the air, and it takes its first breath. Its brain takes over the control of its breathing, stimulated by the fall in the oxygen level in its blood. Its lungs fill with air and its chest expands.

At the same time changes take place in the baby's circulation. These are illustrated in Fig. 11.21. The ductus venosus closes, and all the blood in the hepatic portal vein now has to go through the liver. The ductus arteriosus also closes so that blood in the pulmonary artery can no longer pass into the aorta, and the expanded lungs can now have a good blood supply. The pressure of blood in the right atrium of the heart forces a loose flap of tissue against the foramen ovale, closing the foramen and separating the right side of the heart from the left. All the blood entering the right side of the heart will now be pumped into the lungs, which now take over the work of oxygenation.

The uterine muscles continue contracting after the baby is born. This dislodges the placenta from the uterine wall, and it is expelled – it is known as the *afterbirth*.

Lactation

During pregnancy *oestrogen* causes a system of ducts to grow in the breasts. *Progesterone* promotes the development of glands around the ducts. Look at Fig. 11.22. Find the glands and ducts. Note that the ducts end at the nipple and that there is a layer of fat around the breast.

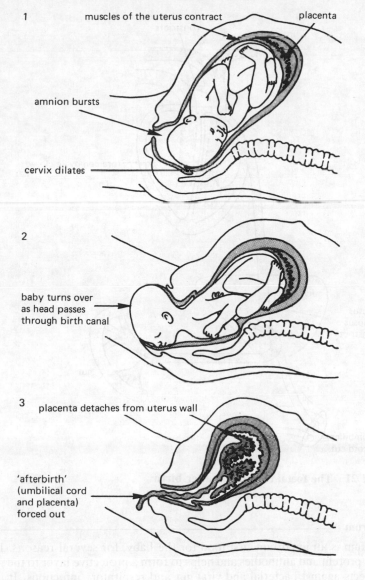

1 muscles of the uterus contract   placenta

amnion bursts

cervix dilates

2

baby turns over
as head passes
through birth canal

3 placenta detaches from uterus wall

'afterbirth'
(umbilical cord
and placenta)
forced out

Figure 11.20   Birth

The glands' production of milk is controlled by the hormone *prolactin*, which is produced by the pituitary gland. They do not produce milk during pregnancy, however, since large amounts of oestrogen inhibit prolactin production.

Once the placenta has been expelled oestrogen levels fall. Prolactin is produced and the glands in the breast start to produce a liquid called *colostrum* and, a few days later, milk. Read the boxed note 'Colostrum'.

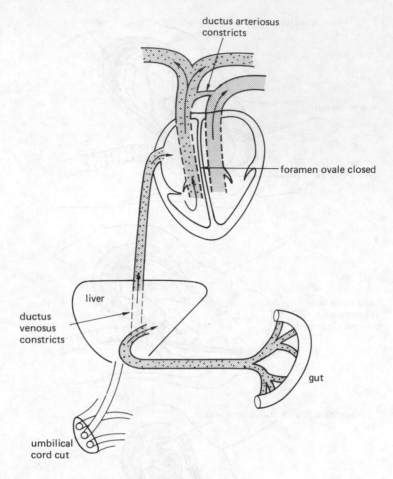

ductus arteriosus
constricts

foramen ovale closed

liver

ductus
venosus
constricts

gut

umbilical
cord cut

Figure 11.21   The foetal circulation after birth

Colostrum

Colostrum is an important first food for the baby, for several reasons. It is high in protein and antibodies and helps to form a protective layer to the gut. It protects against bacterial and viral gut and respiratory infections. It also has a laxative effect, clearing the baby's first stool, the *meconium*.

When the nipple is stimulated by the infant's mouth the pituitary gland begins to release oxytocin – this is a reflex action. The hormone stimulates contractile cells surrounding the glandular cells, causing milk to empty into larger ducts which open on to the nipple. The newborn baby is well adapted for suckling because

- a touch on the cheek will make it turn its head and search for the nipple
- it has a sucking reflex
- it has no teeth.

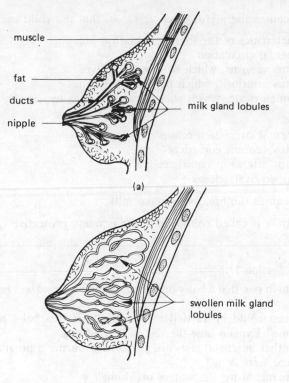

muscle

fat

ducts

nipple

milk gland lobules

(a)

swollen milk gland lobules

(b)

Figure 11.22 Section through a breast: (a) not lactating, (b) lactating

Look at the composition of human milk:

water	lactose	protein	lipids	mineral salts
88.4%	6.5%	1.5%	3.3%	0.3%

Notice that it is quite sweet (lactose is a type of sugar). The composition and volume of human milk change as the baby grows. Even if a baby is born prematurely, the composition of the milk will match what it needs.

Clearly if the composition and volume of the milk change to match the baby's requirements, breastfeeding must be better than bottle-feeding. Other advantages are that the milk is sterile, that it can be given to the baby at any time and that feeding helps to develop a firm emotional attachment between mother and child.

Summary _____

■ During pregnancy women have many *antenatal tests*. These include urine and blood analyses, measurement of weight and blood pressure and estimation of foetal height. Some women also have ultrasound scanning or amniocentesis.

- *Oxytocin* causes the uterus to contract, so that the child can be born.

- The characteristics of the foetal circulation are:
  the placental circulation
  the ductus venosus, which bypasses the liver
  the ductus arteriosis, which bypasses the lungs
  the foramen ovale, which also allows blood to bypass the lungs.

- At birth:
  the placental circulation ceases
  the ductus venosus constricts
  the ductus arteriosus constricts
  the foramen ovale closes.

- *Prolactin* causes the breasts to make milk.

- The first milk is called *colostrum* and has many protective functions.

Questions on Section 11.5

1 There is a high risk that a baby born to a woman over 40 will have Down's syndrome.
  (a) What test could be carried out to find out if a baby has Down's syndrome? Explain how the test is carried out.
  (b) What other antenatal tests are performed during a pregnancy? Why are these tests done?
2 (a) What is meant by 'the waters breaking'?
  (b) What makes the waters break?
  (c) What is the function of these waters during the pregnancy?
3 Not all the blood entering the right side of the foetal heart goes to the lungs.
  (a) How is the blood diverted from the lungs?
  (b) Why doesn't all the blood go to the lungs?
  (c) In some children, the diversions do not properly seal off at birth. What would be the consequences of this for a child?
4 (a) What is colostrum?
  (b) Why is it important that a baby should have the colostrum?
  (c) Give two reasons why breastfeeding a baby is preferable to bottle-feeding.
  (d) Why might it be necessary to bottle-feed a baby?

11.6 Growth and development

Think about the things a newborn baby can do. She can suck, cry, sleep and grip with her hands. She has some vision, though it is limited to objects that are very close. Within a few weeks she can lift her head and smile. During her first year of life she develops quickly. Look at Fig. 11.23. Notice that although a child of six months can hold a toy, she probably cannot feed herself. As she gets older, her co-ordination gets better.

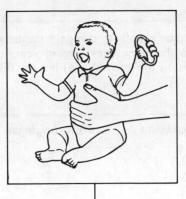

by 6 months

makes lots of noises
sits with support
smiles
holds things and puts
them to the mouth
can roll over

by 9 months

recognises own name
understands quite a lot
can hold a cup and tries
to use a spoon
crawls or shuffles around
may pull himself to his feet
and walk around the furniture

by 1 year

may walk with or
without help

Figure 11.23   Development of a baby in the first year

Growth and development do not go on at the same rate throughout a child's life. Look at Fig. 11.24. Notice that the brain develops most quickly in the first few years of life, while the reproductive organs hardly grow at all until puberty begins at about ten years of age.

Exercise

Look at the table below. Construct line graphs to compare the heights of boys and girls. Put both graphs on the same axes (age in years on the horizontal axis, height in centimetres on the vertical one) but use different colours for the boys and the girls.

Average heights of boys and girls

Age/years	Boys' height/cm	Girls' height/cm
0	52.3	52
1	75	74
2	92	86
3	98	95
4	103	103
5	110	109
6	117	115
7	125	122
8	130	128
9	135	132
10	140	138
11	144	144
12	149	151
13	155	157
14	162	159
15	167	161
16	171	162
17	173	162
18	174	163

What comments can you make about the two graphs you have drawn? Do girls and boys increase in height at the same rate?

Summary

- Babies develop rapidly during their first year.

- Increasing muscular and nervous co-ordination allows a greater range of activity by the baby.

- Rapid growth of the brain and reproductive organs occurs at different times in a person's life.

- Boys and girls grow at different rates.

PROJECT WORK

Spend some time with a baby. Find out what it can do at different ages – at three months and at six months, say. Write down your observations.

What changes in the baby's body have enabled it to develop in the ways that you have observed?

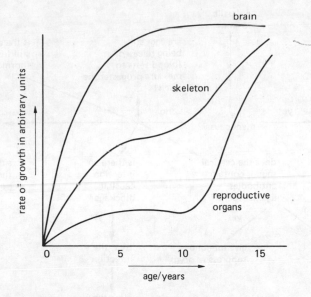

Figure 11.24 **Graph comparing the different rates of growth of some organs in the body**

11.7 IVF and ET

The inability of a couple to have children may be due to many factors. In this section we are going to consider some of the methods that enable infertile women to become pregnant.

If a woman is infertile, causes of the infertility must first be discovered. Then a course of action can be planned. Look at Fig. 11.25. Notice that after determining whether or not the woman is menopausal (that is, if she is still producing egg cells), several questions must be answered. Look at each question carefully and follow the arrows to see what the possible course of treatment is. If pregnancy does not occur even after this checking procedure and treatment, a possible course of action is IVF, which stands for *in vitro fertilisation* (that is, literally fertilisation in a glass vessel) followed by ET which stands for *embryo transplant* (that is, the process whereby the fertilised and 'cultured' egg cell is put back into the woman's uterus).

Let's look at these two processes. For IVF to take place both sperm and egg cells must be available. Harvesting egg cells that are ripe for fertilisation is clearly harder than obtaining the sperm cells. Look at Fig. 11.26. Babies produced as a result of this technique are sometimes called 'test tube babies'.

New methods of finding out when eggs are ready for harvesting are replacing the use of the laparoscope for this purpose. One such method is ultrasound scanning; a second is another scanning technique using nuclear magnetic resonance spectroscopy (NMR). This method gives a good image and is completely safe, but is not widely available yet.

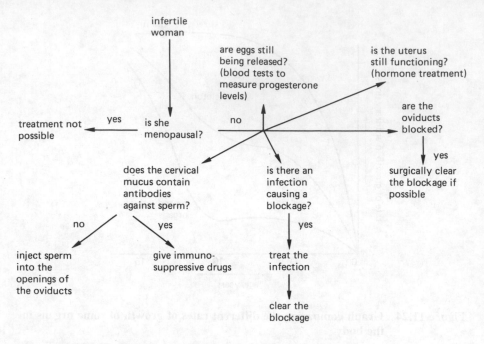

Figure 11.25   Finding a course of action for the infertile woman

Some new techniques reduce the time a woman needs to spend in hospital, from a day or more to just a few hours. For example, eggs in the first stages of fertilisation or zygotes in the two-cell stage can be placed in the egg tube using a long, thin hollow needle.

Exercise

1 Why do the newer techniques try to reduce the time spent in hospital?
2 Why are ultrasound and NMR replacing the laparoscope?

Another method is called Gamete IntraFallopian Transfer or GIFT. Eggs and sperm are placed in the oviducts. This method can use either the mother's own eggs or donor eggs if she is unable to produce them.

Finally, you may have been wondering why many eggs are harvested from the ovaries, rather than just one. All these eggs are mixed with sperm cells, so many embryos are produced. These are used in various ways. Several may be deposited in the uterus – so there may be a multiple birth. Some may be frozen, so that if one embryo does not develop there are more available for transplant.

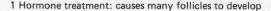

1 Hormone treatment: causes many follicles to develop

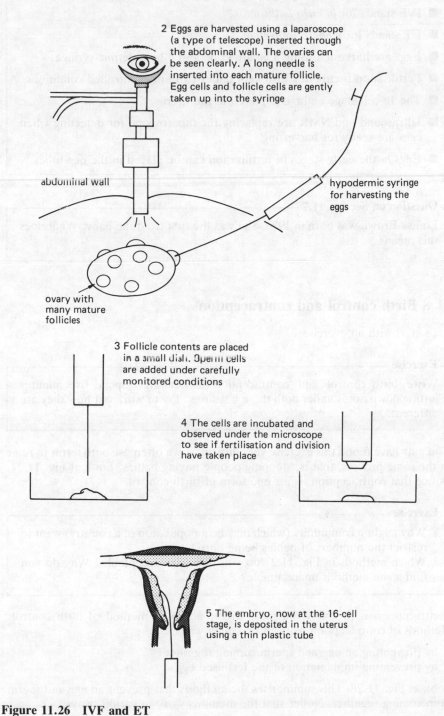

2 Eggs are harvested using a laparoscope (a type of telescope) inserted through the abdominal wall. The ovaries can be seen clearly. A long needle is inserted into each mature follicle. Egg cells and follicle cells are gently taken up into the syringe

abdominal wall

hypodermic syringe for harvesting the eggs

ovary with many mature follicles

3 Follicle contents are placed in a small dish. Sperm cells are added under carefully monitored conditions

4 The cells are incubated and observed under the microscope to see if fertilisation and division have taken place

5 The embryo, now at the 16-cell stage, is deposited in the uterus using a thin plastic tube

Figure 11.26 IVF and ET

┌─ **Summary** ──┐

■ IVF stands for *in vitro fertilisation*.

■ ET stands for *embryo transplant*.

■ Eggs are harvested using a laparoscope and a hypodermic syringe.

■ Fertilisation occurs in a dish kept under carefully controlled conditions.

■ The 16-cell-stage embryo is placed in the uterus.

■ Ultrasound and NMR are replacing the laparoscope for detecting when eggs are ready for harvesting.

■ Eggs in the early stages of fertilisation can be placed in the egg tubes.

└──┘

┌─ **Question on Section 11.7** ──┐

Louise Brown was born in 1978. She was the first test tube baby. What does this mean?

└──┘

11.8 **Birth control and contraception**

Let's start with an exercise.

┌─ **Exercise** ──┐

Write 'birth control' and 'contraception' as headings. Spend five minutes jotting down notes under both these headings. Try to work out how they are different.

└──┘

You may have found this exercise very difficult – we often use both terms to refer to the same process, that is, stopping people having babies. Look at Fig. 11.27. Notice that contraception is just one form of birth control.

┌─ **Exercise** ──┐

1 Why might a community (which may be a population of a country) want to restrict the numbers of babies being born?
2 Which methods in Fig. 11.27 do you think are acceptable? Why do you find some methods unacceptable?

└──┘

Contraception is probably the most widely accepted method of birth control. Methods of contraception work in two main ways:

● by preventing an egg and sperm coming together
● by preventing implantation of the fertilised egg

Look at Fig. 11.28. This summarises the methods that prevent an egg and sperm from coming together. Notice that the methods work in various ways:

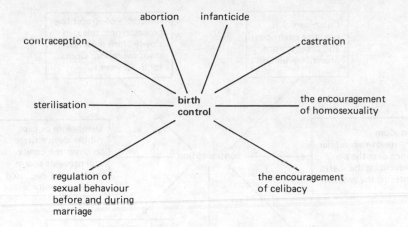

Figure 11.27 Methods of birth control

- some involve the use of a barrier, separating the two types of cell
- some prevent the production of sperm or eggs
- some are permanent, others act for a limited period only
- some rely on abstinence from sexual intercourse or on careful management of its timing, especially when ovulation is most likely.

Look at Fig. 11.28 again. Write down which methods could be included in each of the four groups above.

Now look at Fig. 11.29. This method prevents implantation of the fertilised egg. Why might this form of contraception be unacceptable to some people?

Choice of method depends on effectiveness and acceptability. Look at Table 11.1. Notice that all methods are highly successful if used carefully. So people tend to choose the method they find most acceptable. Sterilisation, for example, may be very effective but is only acceptable to certain people. Some methods have restrictions and disadvantages. Some of these are shown in Fig. 11.30.

Table 11.1 Effectiveness of various contraceptive methods

Method	*Effectiveness/%*
combined pill	99
injectable contraceptives	99
intra-uterine device	96–99
diaphragm (or cap) and spermicide	85–97
sponge	75–91
condom	85–98
natural methods	85–93
female sterilisation	99.6
male sterilisation	99.9

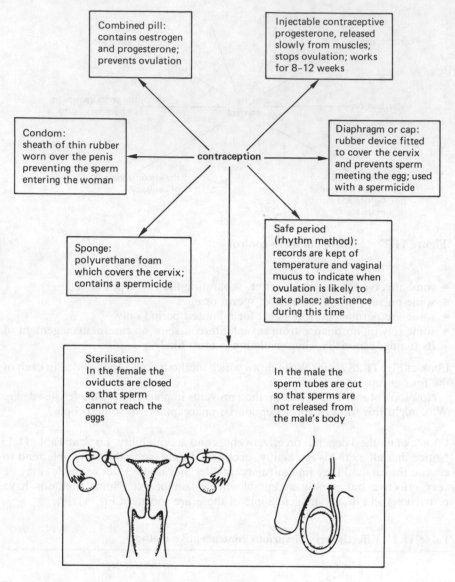

> Combined pill:
> contains oestrogen
> and progesterone;
> prevents ovulation

> Injectable contraceptive
> progesterone, released
> slowly from muscles;
> stops ovulation; works
> for 8–12 weeks

> Condom:
> sheath of thin rubber
> worn over the penis
> preventing the sperm
> entering the woman

contraception

> Diaphragm or cap:
> rubber device fitted
> to cover the cervix
> and prevents sperm
> meeting the egg; used
> with a spermicide

> Sponge:
> polyurethane foam
> which covers the cervix;
> contains a spermicide

> Safe period
> (rhythm method):
> records are kept of
> temperature and vaginal
> mucus to indicate when
> ovulation is likely to
> take place; abstinence
> during this time

> Sterilisation:
> In the female the
> oviducts are closed
> so that sperm
> cannot reach the
> eggs

> In the male the
> sperm tubes are cut
> so that sperms are
> not released from
> the male's body

Figure 11.28 Some methods of contraception

Figure 11.29 Intra-uterine devices (IUDs) like these, usually made of plastic or copper, are fitted into the uterus and prevent implantation

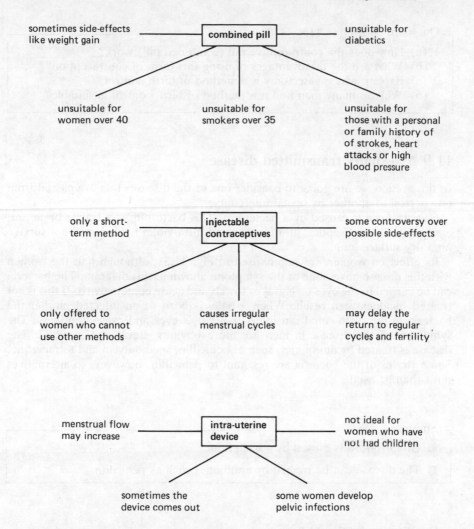

Figure 11.30 Restrictions and disadvantages of some methods of contraception

- **Summary**

■ *Birth control* is a term describing any process that restricts the numbers of babies born.

■ *Contraception* refers to methods of birth control which prevent fertilisation or implantation.

Questions on Section 11.8

1 (a) How does the contraceptive pill (combined pill) work?
   (b) What are the disadvantages of using this form of contraception?
2 (a) Explain why a vasectomy is a method of birth control.
   (b) Why do many men find this method of birth control unsuitable?

11.9 A sexually transmitted disease

In this section we are going to consider one of the diseases that are passed from one person to another by sexual intercourse.

Gonorrohoea is caused by a bacterium. This bacterium is a fragile organism, easily killed by antiseptics, disinfectants and ultra-violet light. It will not survive on a dry surface long.

Its effect on women are summarised in Fig. 11.31, although half the women with this disease have none of the symptoms shown in this diagram. The bacteria can be carried to the eyes on hands or towels and cause conjunctivitis. If this is not treated blindness can result. When a baby is born to an infected mother the bacteria in the birth canal can infect the baby's eyes and cause blindness. The symptoms of the disease in men are more obvious. Look at Fig. 11.32. The disease is treated by antibiotics, such as penicillin, spectomycin and tetracycline. Some strains of the bacteria are resistant to penicillin, however, so alternatives must then be used.

Summary

■ *Gonorrohoea* is caused by a bacterium.

■ The disease can be treated by antibiotics such as penicillin.

Questions on Section 11.9

The use of the condom does not only act as a method of birth control; it can also prevent the spread of disease. Explain this.

PROJECT WORK

Find out about two diseases caused respectively by (a) a virus and (b) a fungus, which can be passed from one person to another in sexual intercourse.

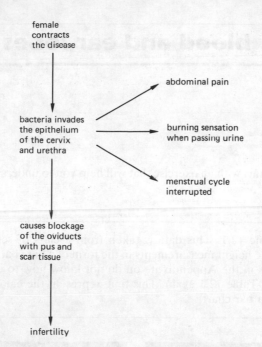

Figure 11.31   The symptoms of gonorrhoea in the female

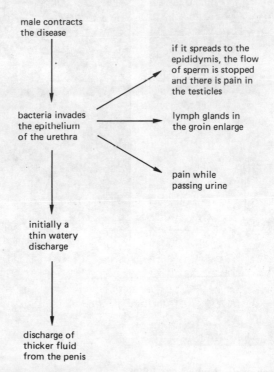

Figure 11.32   The symptoms of gonorrhoea in the male

12.1 Variety

We are going to start with an exercise that will help you to understand how human beings vary.

Exercise

1 Look at Table 12.1. This data is taken from a class of school children. Represent the height measurements in the form of a bar chart (see the note on bar charts in the Appendix if you do not know how to do this).
2 Now look at Table 12.1 again. This time represent the ear lobe results in the form of a bar chart.

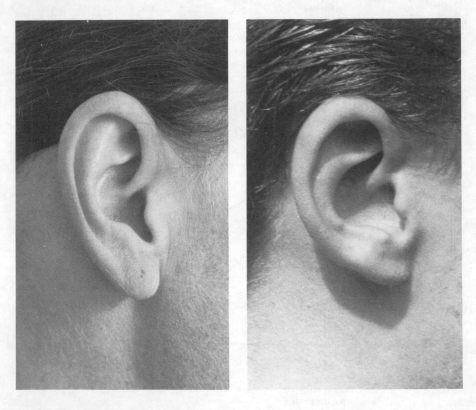

Figure 12.1 Free and attached ear lobes

Table 12.1 Data from a class of school children

Name	Height/cm	Ability to roll tongue	Ear lobes	Blood group
John	115	able	attached	A
Sayarak	130	able	attached	AB
Mike	120	able	free	A
Nazish	118	unable	attached	O
Ruth	100	able	free	O
Ben	121	unable	free	O
Sawako	116	unable	attached	B
Harriet	115	able	attached	A
Ramesh	116	able	free	O

Look at your two charts. They look quite different. Although these represent results from a small group the shapes of the charts would be the same even if they represented results from a large population.

Look at the bar chart representing height. Check the points below with your chart:

- there is great variation in height
- there are few very small children
- there are few very tall children
- most children are in the middle range.

Now let's compare the height chart with the ear lobe bar chart. You will notice that there are only two groups in the ear lobe chart: there is less variation. Look at Table 12.1 once more. Look at the results for tongue rolling and for blood groups. Notice that there is less variation in these characteristics than in height. The results also fall into a definite number of groups.

Exercise

If you were to construct a bar chart to represent blood groups, would it resemble the chart you have drawn for height, or the one for ear lobes?

Characteristics like tongue rolling, ear lobe shapes, blood groups and sex are examples of *discontinuous variation*. Such characteristics are controlled by a single set of genes.

■ **Memory check**
What is a gene?
Where are genes found?
See Section 3.2.

Height is an example of *continuous variation*. It is controlled by many pairs of genes.

┌─ **Exercise** ──┐
│ Can you think of another example of continuous variation? │
└──┘

We all have a particular set of characteristics peculiar to ourselves. In Table 12.1 Sayarak is the only girl with blood group AB who has attached ear lobes and who is 130 cm tall. Individuals can also be identified from their finger prints or from an examination of their DNA (a method known as 'DNA finger printing', although it has nothing to do with fingers!).

┌─ **Exercise** ──┐
│ A person is suspected of a crime. How might DNA 'finger printing' │
│ determine whether he is innocent or guilty? │
│ │
│ [*Clue*: What would a scientist need to do the DNA finger printing?] │
└──┘

We are going to consider another way in which characteristics are expressed. Look at the following examples:

- a seedling kept in the dark is yellow, though it becomes green when it is kept in the light
- a child deprived of vitamin D has malformed bones
- the skin of a person who is normally pale goes brown when exposed to sunlight.

In all of these examples *environmental* factors have had an effect on the organism, that is, its characteristics have been affected by the conditions in which it has lived.

See if you can think of some more examples. In the past scientists have studied identical twins who have been brought up separately. The twins have the same genetic make-up, but they have lived under different environmental conditions. Such studies have helped biologists determine which characteristics are inherited and which are the result of environmental influences. Remember, however, that certain characteristics can only be expressed if the environmental conditions are suitable *and* the appropriate genes are present. For example, a seedling can make chlorophyll only if it is in the light *and* if it has the gene (or genes) for making chlorophyll.

┌─ **EXPERIMENT** ──┐
│ *Aim*: To examine the variation of finger prints and to use finger prints to │
│ identify a person │
└──┘

You will need:
an ink block
plain paper
access to a photocopier (optional)

Method
1 Gently roll your fingers and thumbs of your right hand on an ink pad and press them carefully on a clean sheet of paper.

2 Repeat for your left hand.
3 Identify the types of print, using the photographs in Fig. 12.2.
4 Compare your finger prints with those of your colleagues. You will notice that they are all different. Mark your set of prints with your name. This is your reference set.
5 Either photocopy your prints or do them again; this time the prints should not be marked with the owners' names. Mix up the group's prints. Try to identify the owner of each set of prints by comparing them with the reference sets.
6 Make sure all sets are destroyed and thrown away after the investigation.

Summary

■ Living organisms vary.

■ Variation may be *continuous* in that one characteristic can be exhibited in a wide range of ways.

■ Variation may be *discontinuous* in that one characteristic is exhibited in a limited number of ways.

■ Some characteristics can only be expressed if certain environmental conditions are present.

Questions on Section 12.1

1 Represent the data in Table 12.2 in the form of a bar chart.

Table 12.2 Rhesus factor in a small group of adults

Name	Rh+	Rh−
Ahmad	√	×
Alec	√	×
Dick	√	×
John	√	×
Charles	×	√
Ravindra	√	×
Angela	√	×
Morag	×	√
Pat	√	×
Matora	√	×

2 What sort of variation is represented by your chart?
3 Sketch a bar chart that would represent the variation in height in the adult male population.

| arch | loop | whorl | double whorl |

Figure 12.2   Variation in finger prints

12.2 Setting the scene

We are going to consider how we inherit the information for the characteristics we possess.

■ **Memory check**
Before you go on, make sure that you understand the following terms:

chromosome
homologous chromosomes
gene

Refer to Section 3.2 if you are not sure what these terms mean.

For each inherited characteristic, we have pairs of genes. Each pair is on a *homologous pair* of chromosomes. Look at Fig. 12.3. Notice that:

- the genes for blood group are at the same position on each chromosome; so are the genes for ear lobe shape
- the genes may be the same (look at the genes for blood group)
- the genes may be different (look at the genes for ear lobe shape).

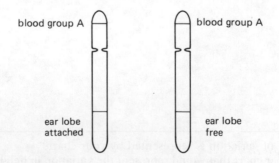

Figure 12.3   Allelic genes

We can call these gene pairs *allelic genes*. Allelic genes are *alternative forms of the same gene*.

Let's consider how a particular pair of genes determines the nature of a characteristic. We'll use eye colour as an example. Of course there's a whole range of eye colours, but to keep our example simple we will say that eye colour can be either blue or brown. So there must be a gene for blue eyes and a gene for brown eyes. Remember that every person must have two genes governing eye colour. Studies have shown that people who have two brown-eye genes all have brown eyes. People who have two blue-eye genes have blue eyes, and people who have one blue-eye gene and one brown-eye gene have brown eyes.

These brown-eyed people with one blue-eye gene and one brown-eye gene are interesting. Their blue-eye gene is *masked* by the brown-eye gene. They can pass on the blue-eye gene to their children but it is not expressed in themselves. Genes that can be masked are called *recessive* genes. Genes that can mask another gene are *dominant*.

Before we go on let's represent these genes more simply, using letters. The actual letter we use is unimportant, but we have to be very careful in the use of capital and lower-case (small) letters. Let's see how the system works. We'll call the gene for eye colour B.

- we write the gene for brown eyes as B (it's a dominant gene so we use a capital letter)
- we write the gene for blue eyes as b (it's a recessive gene so we use a lower-case letter).

The genes a person has for a particular characteristic are called the *genotype*. The appearance of the person (the expression of these genes) is called the *phenotype*. Thus in a person with two brown-eye genes (BB), who therefore has brown eyes,

- BB = genotype
- brown eyes = phenotype.

There are three different genotypes for eye colour in our example: BB, Bb and bb.

Look at the genotypes BB and bb first. They have the same genes. We call these genotypes *homozygous*. Now look at the genotype Bb. This has different genes. We call genotypes like this *heterozygous*.

It would be a good idea to read over this section again and learn those terms that are new to you.

Summary

■ *Allelic genes* are alternative forms of the same gene.

■ Letters are used to represent genes.

■ Some genes can mask other genes. They are *dominant* and are represented by capital letters.

■ Those genes that can be masked by dominant genes are *recessive* and are represented by lower-case letters.

■ The term *phenotype* refers to the appearance of the person or to the expression of the genes.

■ The term *genotype* refers to the genes an individual has.

■ Genotype can be *homozygous* or *heterozygous*.

Questions on Section 12.2

1 The ability to roll the tongue is controlled by a dominant gene.
  (a) What letter could you use to represent the tongue-rolling gene?
  (b) What letter could you use to represent the recessive gene for not being able to roll the tongue?
  (c) What is the genotype for a non-tongue-roller?
  (d) What are the two possible genotypes for a tongue-roller?
2 A person has the genotype BB for eye colour. He has brown eyes.
  (a) Is he homozygous or heterozygous for eye colour?
  (b) What is his phenotype?

12.3 Cell division with a difference

In Section 11.1 you learnt that human beings reproduce sexually, and that they produce special cells called *gametes* which fuse together to produce a *zygote*.

■ **Memory check**
What are these gametes called?
Where are they produced?
See Sections 11.1, 11.2 and 11.3.

In this section we are going to look at how these gametes are produced and what makes these gametes so special.

You probably remember from Section 3.2 that human body cells contain 46 chromosomes, or 23 pairs of homologous chromosomes. The sperm and egg cells each contain just 23 chromosomes, however. So when two gametes fuse a cell with 46 chromosomes is formed.

Cell division that results in cells containing half the normal body cell number of chromosomes is called *meiosis*. Clearly meiosis will only occur where the eggs and sperm cells are made, that is, in the ovaries and testes.

Let us consider a homologous pair of chromosomes. Look at Fig. 12.4(a). Notice that one chromosome is of maternal origin. This means it came from the egg cell that made this person. The other chromosome is of paternal origin. This chromosome came from the sperm that made this person. When the cell divides by meiosis, one chromosome of each pair will go into the daughter cells. We have drawn one pair of allelic genes (marked Aa) on these chromosomes. Look at Fig. 12.4(b). Notice that:

• a gamete will contain *one* gene for each characteristic
• one chromosome of each homologous pair will go into each gamete.

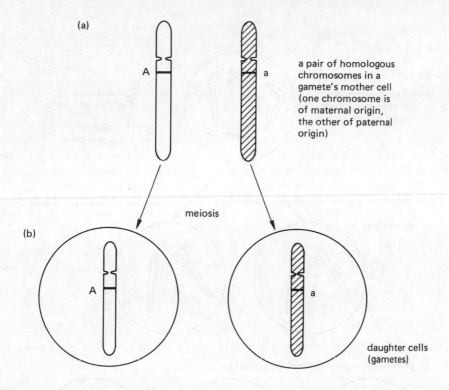

Figure 12.4   Separation of a pair of homologous chromosomes during meiosis

Let's think about another example of this. Look at Fig. 12.5(a). This cell contains six chromosomes and not 46. There are three homologous pairs of chromosomes in this cell. Gametes derived from this cell will only have three chromosomes each (one of each pair). Look at Fig. 12.5(b) to check this.

But the combination of chromosomes shown in Fig. 12.5(b) is not the only possibility. Look at Fig. 12.5(c). It shows that there are many possible combinations of the six chromosomes in Fig. 12.5(a). Human cells have 46 chromosomes, so the number of possible combinations of maternal and paternal chromosomes is enormous. Notice from Fig. 12.5(c) that the combinations of maternal and paternal chromosomes need not have existed before. This means that individuals resulting from these gametes may vary from their parents, possibly with characteristics that are to the individual's advantage.

Finally, when a cell divides by meiosis it does not simply divide into two. Look at Fig. 12.6. Notice that in the first stage of meiosis (as in mitosis) the chromosomes duplicate. The cell then divides and divides again, giving four gametes each with 23 chromosomes.

Meiosis is a very special form of cell division, and it is important to know and understand the process. In particular, the potential for change and variation arising from new chromosome combinations produced in meiosis provides a basis for evolution – you will find out more about evolution in Section 12.8.

This is also a good place to revise mitosis! (See Section 3.2.)

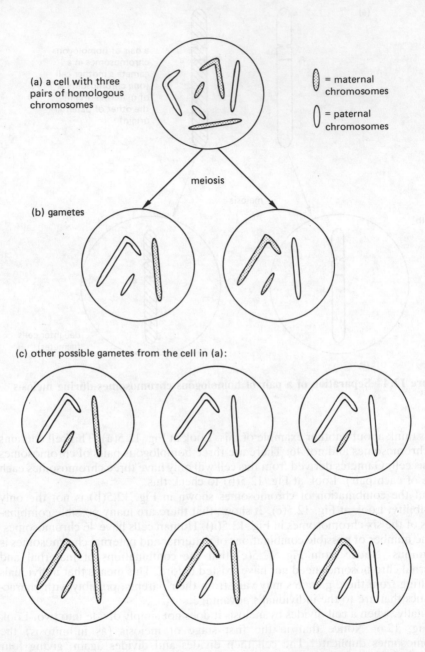

(a) a cell with three pairs of homologous chromosomes

= maternal chromosomes

= paternal chromosomes

meiosis

(b) gametes

(c) other possible gametes from the cell in (a):

Figure 12.5   Separation of several pairs of homologous chromosomes during meiosis

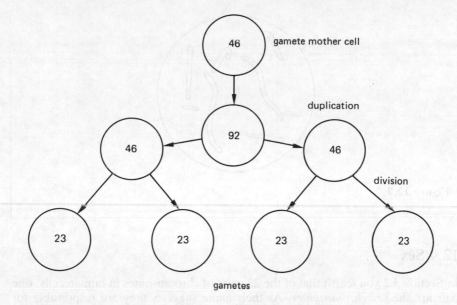

Figure 12.6   Numbers of chromosomes produced during meiosis

Summary

- Meiosis results in the production of sperm and egg cells. These are gametes.

- Meiosis occurs in the testes and ovaries.

- Gametes have 23 chromosomes. Human body cells have 46 chromosomes.

- Each gamete contains one chromosome from each pair of homologous chromosomes. There may be a mixture of maternal and paternal chromosomes.

- Mixing of maternal and paternal chromosomes results in material for variety and change in the offspring.

Questions on Section 12.3

1 Look at the cell in Fig. 12.7.
   (a) How many chromosomes are in this cell?
   (b) How many homologous pairs of chromosomes are in this cell?
   (c) Draw two cells that could be produced by this cell as a result of meiosis.
   (d) If this cell divided by mitosis, how many chromosomes would there be in each resulting cell?

2 (a) Where does meiosis take place in the human body?
   (b) How does meiosis provide material for change?

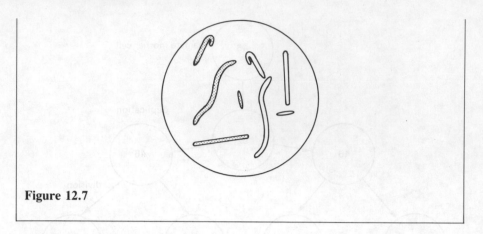

Figure 12.7

12.4 Sex

In Section 3.2 you learnt that of the 23 pairs of chromosomes in human cells, one pair are the *sex chromosomes*. As their name suggests they are responsible for determining the sex of the individual.

Examination of sex chromosomes of males and females reveals that there are two types of sex chromosomes which are known as X and Y. Look at Fig. 12.8. Notice that the Y chromosome is much smaller than the X chromosome. Notice also that female cells have two X chromosomes (XX) and male cells have one X and one Y chromosome (XY).

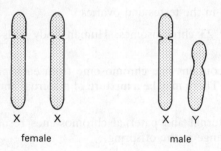

X   X   X Y

female   male

Figure 12.8 Sex chromosomes

Sex chromosomes are a *homologous pair* of chromosomes (refer back to Sections 3.2 and 12.3).

When gametes are formed (as a result of meiosis), homologous chromosomes separate. Therefore in each gamete there will be *one* sex chromosome. Look at Fig. 12.9. Notice that

- *all* egg cells carry an X chromosome
- 50% of sperm cells carry a Y chromosome
- 50% of sperm cells carry an X chromosome.

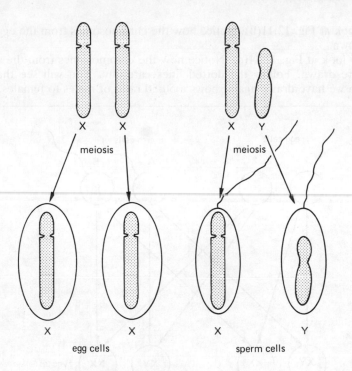

Figure 12.9 Separation of sex chromosomes at meiosis (not drawn to scale)

Each time an egg is fertilised it combines with a sperm. It is equally likely to combine with either a Y-carrying sperm or an X-carrying sperm. So each time an egg is fertilised it could result in a male (XY) or a female (XX). Statistically, there is an equal chance that the new individual will be male or female.

Look at Fig. 12.10, which demonstrates this probability. Follow the leader lines from the egg cell 1 to the zygotes 1 and 2. The lines show where the X chromosome from these gametes will go. This gamete is equally likely to fuse with sperm A or sperm B. If it fuses with the sperm A it forms zygote 1, and if it fuses with sperm B it forms zygote 2.

Now follow the leader lines from egg cell 2 to zygotes 3 and 4. Egg cell 2 is equally likely to fuse with sperm A or with sperm B, forming either zygote 3 or zygote 4.

Go over the last two paragraphs again – make sure you understand Fig. 12.10 before you go on.

Another way of demonstrating probabilities is described in the boxed note 'Probability and the chequerboard'.

Probability and the chequerboard

Look at Fig. 12.11(a). This method of demonstrating probability is called the *chequerboard*. It deals with a single pair of chromosomes. Notice that

- two possible gametes from the male are written down one side of the grid
- two gametes from the female are written along the top of the grid.

Now look at Fig. 12.11(b). Notice how the chromosomes from the egg cells are drawn.

Now look at Fig. 12.11(c). Notice how the chromosomes from the sperm cells are drawn. Follow the dotted lines carefully. You will see that the pattern we have drawn again shows a 50:50 ratio of males to females.

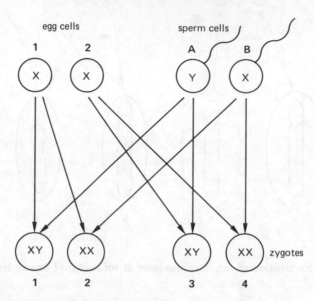

Figure 12.10   Finding the probability of male and female children

Exercise

The zygotes in Fig. 12.10 show the genotypes of the offspring. Can you say which of the zygotes 1, 2, 3 and 4 are males and which are females? (That is, what are their phenotypes?)

Fig. 12.10 shows all the possible combinations of these gametes. What is the probability of a child being female?

Summary

■ There are X and Y *sex chromosomes*.

■ A female has two X chromosomes in her cells. A male has an X and a Y chromosome in his cells.

■ Gametes contain one sex chromosome.

■ All egg cells contain X chromosomes and can fuse with a sperm carrying either an X or a Y chromosome.

■ There is a 50:50 chance that a child will be a boy or a girl.

■ There are two ways of demonstrating this probability.

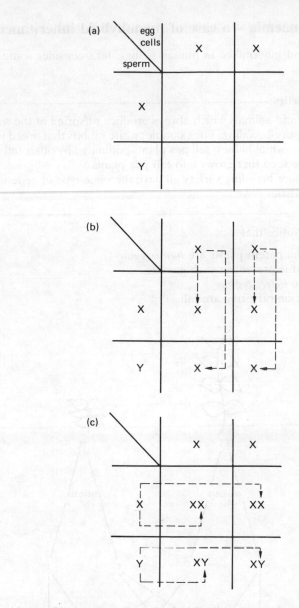

Figure 12.11 Chequerboard method of finding the probability of male and female children

There is a 50:50 chance of a child being a male or a female. Explain why this is so by means of diagrams and notes.

12.5 **Sickle cell anaemia – a case of monohybrid inheritance**

To help us understand inheritance in human beings, let's consider some pure-breeding plants.

Pure-breeding varieties

Varieties of plants and animals which always produce offspring of the same type are said to be pure-breeding. For example, white rabbits that breed with each other only have white babies; tall pea plants pollinated by other tall pea plants only produce seed that grows into tall pea plants.

Organisms of a pure-breeding variety all have the same type of gene for a particular characteristic.

Look at Fig. 12.12. Notice that

- the genotypes of the parent plants are *homozygous*
- one gene for height is present in each gamete
- all the offspring are *heterozygous*
- all the first-generation offspring are tall.

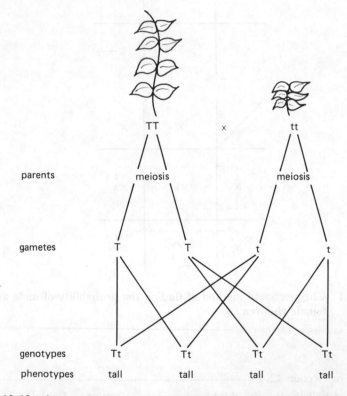

Figure 12.12   A cross between tall and dwarf pure-breeding plants; the symbol F₁ refers to the first filial generation, that is, the offspring from two different pure-breeding organisms

So tallness is the dominant gene and dwarfness the recessive gene. Can you see why? Go back to Section 12.3 if you are not sure about this.

Exercise

Look at Fig. 12.12. Represent the cross shown in this figure in a chequer-board. (Refer back to Section 12.4 if necessary.)

Now let's see what will happen if we cross two F_1 plants. Look at Fig. 12.13. Notice that:

- the genotypes of the parents are heterozygous
- one gene for height is present in each gamete
- the offspring (F_2) are a mixture of tall and dwarf plants in a 3:1 ratio
- the genotypes of the offspring (F_2) can be homozygous or heterozygous.

Exercise

The cross in Fig. 12.13 is represented as a chequerboard. Write the cross out using the leader lines as in Fig. 12.12.

The 3:1 ratio indicates the inheritance of a characteristic that is controlled by one gene. This is called *monohybrid inheritance*. Human beings are not suitable organisms for genetic experiments, partly because their life cycle is so long and

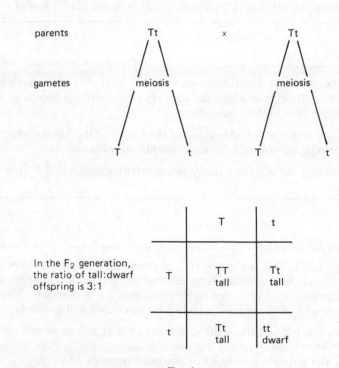

Figure 12.13 A cross between two F_1 plants

because they do not possess many characteristics that are controlled by just one gene, but mainly because most people think it would be wrong to carry out experiments on breeding their fellow-humans. Insects like fruit flies or flour beetles are usually used because they have short life cycles and they have many characteristics that are each controlled by one gene.

Sickle cell anaemia in human beings is controlled by one gene. The red blood cells in a person with sickle cell anaemia are mis-shaped and cannot carry oxygen as well as normal red blood cells. We can write the gene for normal haemoglobin as Hb^A. The gene for sickle cell anaemia is Hb^S. Look at the following genotypes for sickle cell anaemia:

$Hb^A Hb^A$ = normal
$Hb^A Hb^S$ = normal (but heterozygous)
$Hb^S Hb^S$ = sickle cell anaemia

The normal Hb^A gene is not completely dominant over the Hb^S gene, however. This has interesting consequences for people who are heterozygous. You will learn more about this in Section 12.8.

--- Exercise --

Two people with the genotypes $Hb^A Hb^S$ and $Hb^A Hb^S$ have a child. What is the probability that this child will have sickle cell anaemia?

[*Clue*: First work out the possible gametes that the parents can produce. Then complete the cross with a chequerboard or using leader lines.]

--- Summary --

■ A cross between two individuals who are heterozygous for a given gene will result in offspring in which the relevant characteristics appear in a 3:1 ratio. This is *monohybrid inheritance*.

■ For reasons of morality and because of the length of the human life cycle, human beings are not used for experiments on inheritance.

■ Human beings do not have many characteristics controlled by just one gene.

--- Questions on Section 12.5 --

Look at Fig. 12.14. This is a family tree or pedigree. The squares represent males and the circles represent females. Notice that both daughters have normal haemoglobin. The first daughter has married a man who also has normal haemoglobin. They have a child who has sickle cell anaemia.

1 What does this tell you about the genotypes of daughter number 1 and her husband?
2 What are the possible genotypes of daughter number 2?

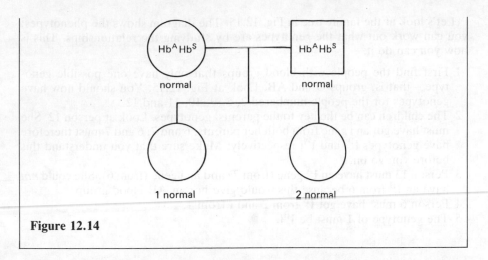

Figure 12.14

12.6 **Blood**

In Section 6.5 you learnt that there are four different blood groups: A, B, AB and O. In this section we are going to examine how these different groups are inherited.

In the population, three genes are responsible for determining one's blood group. Each individual of course has only two of these genes. The genes determining blood group are written I^A, I^B and i (or I^O). (Note that they are written in a slightly different way from genes representing eye colour or tallness of pea plants.) Combinations of different pairs of these genes determine which blood group a person will have. Look at Table 12.3. Notice that

- blood group is the *phenotype*
- I^A must be dominant to i
- I^B must be dominant to i
- i is therefore a recessive gene
- I^A and I^B are equally dominant. They are said to be *co-dominant*.

Remember that the two genes determining a person's blood group come from his or her parents. So a child with blood group AB, whose genotype is $I^A I^B$, got the gene I^A from one parent and the gene I^B from the other.

Table 12.3 Phenotypes and genotypes of blood groups

Blood group (phenotype)	Genotype
A	$I^A I^A$ or $I^A i$
B	$I^B I^B$ or $I^B i$
AB	$I^A I^B$
O	ii

Let's look at the family tree in Fig. 12.15. The diagram shows the phenotypes; you can work out what the genotypes are by studying the relationships. This is how you can do it.

1 First find the people with blood groups that only have one possible genotype – that is, groups O and AB. Look at Fig. 12.15. You should now have genotypes for the people numbered 2, 5, 9, 10, 11 and 12.

2 The children can be the key to the parents' genotypes. Look at person 12. She must have got an i gene from both her parents, 6 and 7; 6 and 7 must therefore have genotypes $I^B i$ and $I^A i$ respectively. Make sure that you understand this before you go on.

3 Person 13 must have an I^A gene (from 7) and an i gene (from 6). She could *not* have an I^B from 6 because this would give her an AB blood group.

4 Person 6 must have got I^B from 2 and i from 1.

5 The genotype of 1 must be $I^B i$.

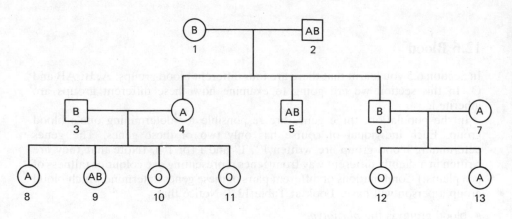

Figure 12.15   Family tree – blood groups

Exercise

What are the genotypes of 3 and 4 in the family tree in Fig. 12.15?
  Remember to start with the genotypes for O and AB groups, and begin with the children 8, 9, 10 and 11.

Summary

■ The ABO system of blood grouping is controlled by three genes I^A, I^B and i.

■ I^A and I^B are co-dominant.

■ I^A and I^B are dominant to i.

Questions on Section 12.6

1 Find the correct parents for the children below:

| Sophie | AB | Saleem | B |
| Romesh | A | Amy | O |

Blood groups of the parents are:

Mr Drayton	O	Mr Parke	O
Mrs Drayton	O	Mrs Parke	AB
Mr Prakesh	A	Mr Mais	A
Mrs Prakcsh	O	Mrs Mais	B

12.7 Mistakes

Sometimes there is a change in a gene or part of a chromosome which causes a change in the characteristic it controls. If the change occurs in a gamete or zygote the entire organism can be affected. This change may then be passed on to future generations. Changes in chromosomes are called *mutations*.

Most mutations are recessive and therefore are not apparent in phenotypes. Look at Fig. 12.16. Notice that 50% of the children can carry the mutated gene. They have normal characteristics, however, because they have a normal dominant gene to mask the mutation.

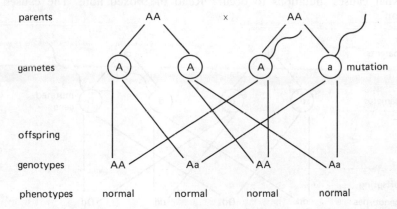

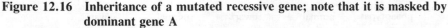

Figure 12.16   Inheritance of a mutated recessive gene; note that it is masked by dominant gene A

Exercise

1 Some of the offspring in Fig. 12.16 will produce gametes with the mutated 'a' gene. If one of these offspring has children with a person whose genotype for this characteristic is AA, what will the genotypes and phenotypes be?

2 If a person with genotype Aa has a child with a person whose genotype is also Aa, what is the likelihood that their child will not be normal for this characteristic?

From this exercise you can see that it is only when two recessive mutated genes come together that the abnormality is shown in the phenotype.

Most mutations are harmful to the organism. Some can be advantageous, however, and these may lead to the success of the individual that has the mutation. So mutations can form the basis of evolutionary change. Environmental factors may *select* mutants with advantageous characteristics – you will learn more about this in the next section. Most mutations are rare, however, and rare recessive mutations are unlikely to be expressed in the phenotypes of the children. If you do not understand why this is so, try the exercise below.

Exercise

A normal genotype is AA. If one of the genes mutates, the genotype becomes Aa. Phenotypes showing the abnormality must have genotype aa.

1 A person has genotype Aa. As the mutation is rare all partners are likely to be AA. What will the genotypes of their offspring be?
2 Will any of the offspring show the abnormality? Explain your answer.

One kind of dwarfism is a dominant mutation affecting one person in 20 000. This is more likely to be expressed in the phenotypes. Look at Fig. 12.17, where D is the dominant dwarfism gene and d the recessive normal gene. Notice that there is a 50% chance of a child being a dwarf if a gamete has this mutation.

So what causes mutations to occur? Read the boxed note 'The causes of mutation'.

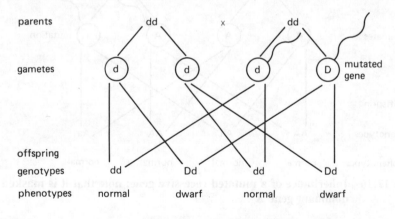

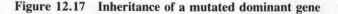

Figure 12.17   Inheritance of a mutated dominant gene

The causes of mutation

Some mutations are spontaneous – not related to the environment. But chromosomes are made up of DNA (see Section 3.2); DNA is very stable, and rarely changes spontaneously.

Other mutations are produced by *mutagenic agents* in the environment. These are factors that increase the rate of mutation. They include X-rays, gamma-rays, ultra-violet light and certain chemicals, such as mustard gas.

── Summary ──────────────────────────────

■ A *mutation* is a change in a gene or part of a chromosome.

■ Mutations that affect the whole individual and that can be passed to its offspring occur in the gamete or the newly formed zygote.

■ Mutations can be spontaneous or induced by mutagenic agents.

■ Most mutations are rare and recessive.

■ Mutations can form the basis of evolution.

── Questions on Section 12.7 ──────────────────

1 What is a mutation?
2 When can a mutation take place?
3 What are the causes of mutations?

12.8 Changing, selecting and evolving

Let's start this section by looking again at the inheritance of sickle cell anaemia, which we met in Section 12.5.

Look at Fig. 12.18. Notice that the person heterozygous for the condition has an advantage in areas where malaria is common. Environmental conditions have worked to *select* the sickle cell gene. Such selection is called *natural selection*. The selection results in keeping the gene in the population.

── Exercise ──────────────────────────────

1 Two people have the same genotype, $Hb^A Hb^S$. What are the possible genotypes of their children?
2 Will environmental conditions select the Hb^S gene in areas where malaria is absent? Explain your answer.

Now look at Fig. 12.19. Notice that:

● the plants produce more offspring than can survive
● individuals are forced to compete for space, light and water
● the organisms show a great deal of variation
● the plants that are more successful in getting light, water and space will survive at the expense of the slower-growing plants
● the larger, quicker-growing plants pass their genes on to their offspring.

But if conditions changed so that the slower-growing plants were more successful, these plants would be the ones to be selected. Over many generations, selection

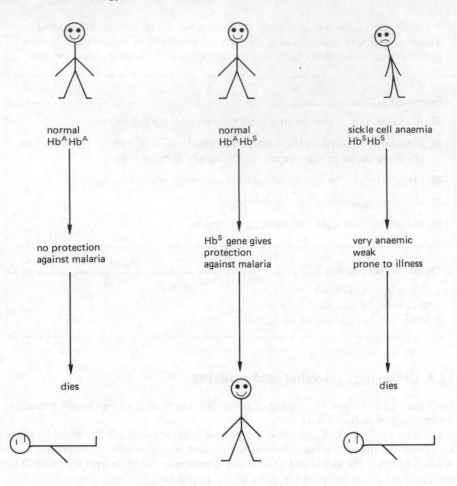

normal
Hb^{A}Hb^A

normal
Hb^{A}Hb^S

sickle cell anaemia
Hb^{S}Hb^S

no protection
against malaria

Hb^S gene gives
protection
against malaria

very anaemic
weak
prone to illness

dies

dies

Figure 12.18   Natural selection and sickle cell anaemia

can lead to the disappearance of some genes. The organism, and in time even the species, may change. This is how organisms evolve. Fossil evidence suggests that horses evolved from animals that had all their toes on the ground. Look at Fig. 12.20, which shows the leg bones of fossil horses and those of today. Notice the gradual change to the bones of the feet.

Exercise

What are the advantages to the horse in having feet like those in Fig. 12.20(d)? Why was the animal having these feet selected in preference to those with feet like those in Fig. 12.20(a)?

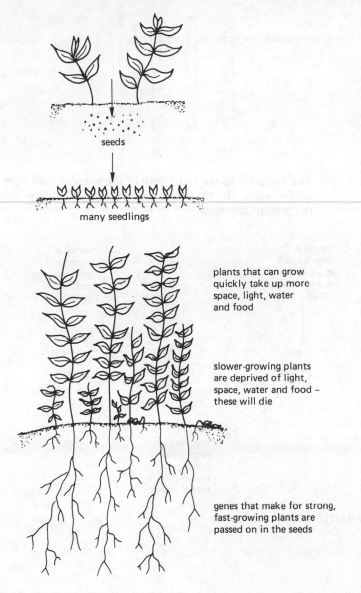

plants that can grow
quickly take up more
space, light, water
and food

slower-growing plants
are deprived of light,
space, water and food –
these will die

genes that make for strong,
fast-growing plants are
passed on in the seeds

seeds

many seedlings

Figure 12.19   Natural selection

Selection is not always natural, however. Consider the cultivated and wild
strawberry. Cultivated strawberries give much larger yields of much larger berries
than the wild plants do. Such plants are the result of *artificial selection*. Most of
our fruit, vegetables, cereals and meat are the product of artificial selection.

Breeders of plants and animals select organisms with characteristics that they
want. Gametes from these organisms are brought together so that the offspring
will have a combination of desired characteristics. Look at Fig. 12.21.

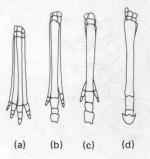

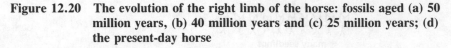

(a)　　(b)　　(c)　　(d)

Figure 12.20　The evolution of the right limb of the horse: fossils aged (a) 50 million years, (b) 40 million years and (c) 25 million years; (d) the present-day horse

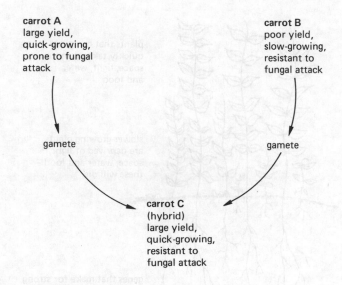

carrot A
large yield,
quick-growing,
prone to fungal
attack

carrot B
poor yield,
slow-growing,
resistant to
fungal attack

gamete

gamete

carrot C
(hybrid)
large yield,
quick-growing,
resistant to
fungal attack

Figure 12.21　Artificial selection

Exercise

Why are plant and animal breeders concerned that older breeds of the organism should be preserved?

Summary

■ Organisms show *variation* within a species.

■ Organisms produce more offspring than can survive.

■ Environmental conditions will select organisms that are successful.

■ Successful organisms will pass on their genes.

■ *Natural selection* leads to evolution.

■ Desirable characteristics can be combined in an organism. This is *artificial selection*.

Questions on Section 12.8

1 A variegated plant is never found growing wild. Why is this?
2 The 'greenhouse effect' is expected to cause rapid and dramatic changes in the climate. Some species of plants and animals are expected to die out because they will not be able to evolve fast enought to adapt to these changes. What does this mean?

12.9 Using the system: genetic engineering

Before you start this section make sure that you understand the following points:

• information for characteristics that are inherited is carried on chromosomes
• each characteristic is controlled by one or more genes
• the genes on the chromosomes are like beads on a string
• chromosomes are made of DNA.

Understanding how DNA works has helped scientists to design new techniques to extract genes for known characteristics and introduce them into new cells. Initially the genes were transferred between different strains of bacteria or between mammalian cells. The exciting discovery was that the transferred genes worked, and could be inherited – that is, if the 'infected' cell divided the daughter cells inherited the new gene.

It was also found that if a gene controlling the production of a growth hormone was transferred into a mouse egg cell, the mouse grew much larger than normal. Its offspring were also larger. The transfer of growth hormone genes has been used to increase the rates of growth or the final size of pigs and sheep.

In Section 3.2, you saw that every gene has a particular position on a chromosome. A technique called *homologous recombination* directs genes to particular locations in order to repair genetic defects. For example, when portions of haemoglobin genes are added to cultures of human cells they become correctly located. The technique uses an organism called a *retrovirus*. Normally retroviruses invade cells and cause diseases. But if the disease-causing genes are removed from the retrovirus and replaced by therapeutic genes, the virus will invade the cell, attach itself to the chromosome and repair the genetic defect. Look at Fig. 12.22. Notice that the therapeutic gene does not need to replace the faulty gene. Using a virus to take in the DNA is called *transfection*.

The technique may prove useful in the treatment of single-gene disorders using body cells. For example, it could be used to treat disorders in haemoglobin production. Look at Fig. 12.23. The possibility of treating genetic defects in humans in this way is very exciting. But there are problems.

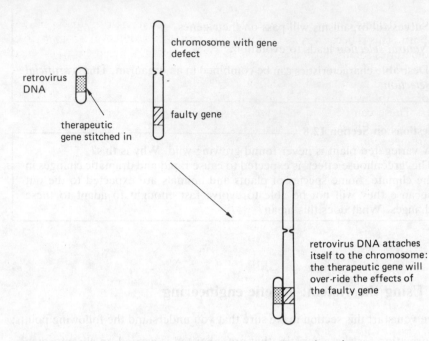

Figure 12.22   Repairing genetic defects by genetic engineering

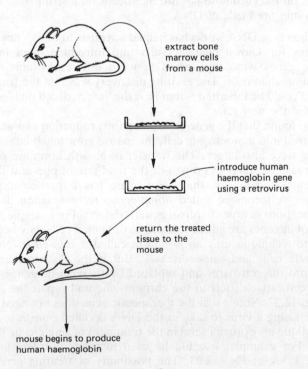

extract bone
marrow cells
from a mouse

introduce human
haemoglobin gene
using a retrovirus

return the treated
tissue to the
mouse

mouse begins to produce
human haemoglobin

Figure 12.23   Gene therapy by gene transfer

Genetic defects can only be repaired by gene transfer after an organism has been born. This is because many embryos die; in addition the results are variable, and sometimes totally unpredictable. Genes that took up a wrong position might interfere with other normal genes on the chromosome; some scientists also fear that they could cause a cell to become cancerous.

Another application of recombinant DNA technology in the use of DNA probes. These can be used to detect abnormal genes in embryos, foetuses and adults. Look at Fig. 12.24. The test is highly specific and reliable, and only a few cells are needed.

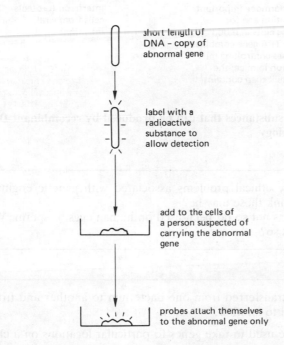

short length of DNA - copy of abnormal gene

label with a radioactive substance to allow detection

add to the cells of a person suspected of carrying the abnormal gene

probes attach themselves to the abnormal gene only

Figure 12.24 Diagnosis using a DNA probe: detecting an abnormal gene

The realisation that everyone's genetic make-up is slightly different (except for identical twins) has led to the development of genetic finger printing, mentioned in Section 12.1. It is an invaluable tool for identifying criminals, and children who have become separated from their parents.

Finally, recombinant DNA technology is used in the production of hormones and other biologically active substances. Fig. 12.25 gives a summary of some of the substances that can be made in this way.

Genetic engineering is in its infancy. We are not yet fully aware of all the possibilities that it will offer. For example, the study of the locations of genes on human chromosomes may help us to determine whether a person is likely to develop coronary heart disease or diabetes.

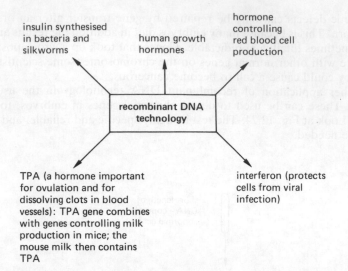

insulin synthesised
in bacteria and
silkworms

human growth
hormones

hormone
controlling
red blood cell
production

recombinant DNA
technology

TPA (a hormone important
for ovulation and for
dissolving clots in blood
vessels): TPA gene combines
with genes controlling milk
production in mice; the
mouse milk then contains
TPA

interferon (protects
cells from viral
infection)

Figure 12.25   Some substances that can be produced by recombinant DNA technology

---

Exercise

1 There are many ethical problems associated with genetic engineering. What do you think these may be?
2 Gene therapy has not so far been used on human eggs or sperm. Why do you think this is so?

Summary

■ Genes can be transferred from one bacterium to another and from one mammalian cell to another.

■ *Retroviruses* are used to take genes to particular locations on a chromosome.

■ Genetic defects can be repaired by *gene transfer*.

■ *DNA probes* can be used to detect abnormal genes.

■ *Recombinant DNA technology* can also be used to produce hormones and other biologically active substances.

■ *Genetic finger printing* is based on the fact that everyone's DNA map is unique.

Questions on Section 12.9

1 Give an example of genetic engineering.
2 How can genetic engineering help to cure some people of diseases?

13 Understanding the good and the bad

13.1 What are micro-organisms?

Micro-organisms, or *microbes*, are organisms that are too small to be seen with the naked eye. You can only see them with a microscope or when there are large numbers together.

There are many types of micro-organism. Let's look at a few examples.

Viruses

These are very, very small and have simple geometric shapes. They do not respire, feed or grow unless they are inside a living cell. They are therefore *parasites* (read the boxed note 'Parasites'). Look at Fig. 13.1.

Parasites

A parasite lives on another living organism, which becomes its *host*. The host does not gain anything from the relationship. Sometimes the parasite makes its host ill or kills it.

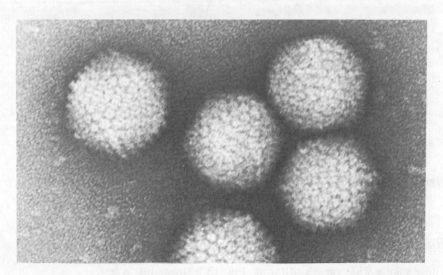

Figure 13.1 **Virus particles, magnified hundreds of thousands of times. This virus causes the symptoms of the common cold**

Fungi

This group includes mushrooms, toadstools, moulds and yeasts. They need warmth, food and moisture to grow. Look at Fig. 13.2. Notice that yeasts reproduce by budding, and that filamentous fungi produce spores. Some fungi live on dead organic material and others on living organisms.

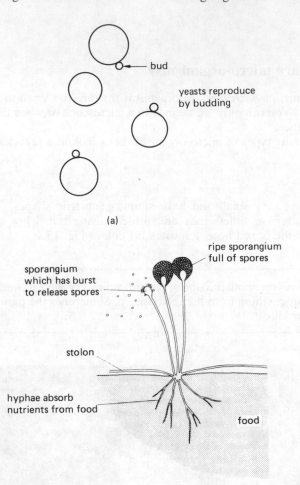

bud

yeasts reproduce
by budding

(a)

ripe sporangium
full of spores

sporangium
which has burst
to release spores

stolon

hyphae absorb
nutrients from food

food

(b)

**Figure 13.2   Two types of fungus: (a) yeasts, (b) filamentous fungus – this is
the pin-mould fungus (*Rhizopus*) often seen on stale bread**

Bacteria

These take a wide variety of forms. Fig. 13.3 shows just a few.

Bacteria can exist in two forms: in the *vegetative* state and as *spores*. When bacteria are in the vegetative state they can divide by *binary fission*. That is, they

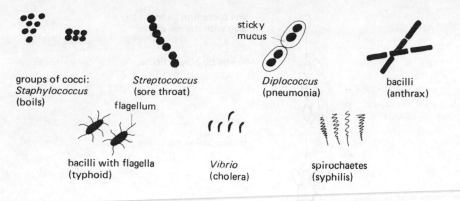

groups of cocci:
Staphylococcus
(boils)

Streptococcus
(sore throat)

flagellum

sticky
mucus

Diplococcus
(pneumonia)

bacilli
(anthrax)

bacilli with flagella
(typhoid)

Vibrio
(cholera)

spirochaetes
(syphilis)

Figure 13.3   Different types of bacteria

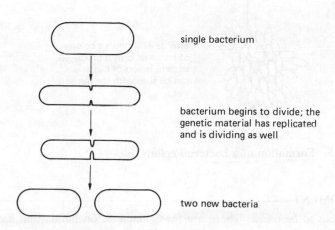

single bacterium

bacterium begins to divide; the
genetic material has replicated
and is dividing as well

two new bacteria

Figure 13.4   Binary fission in bacteria

split into two. Look at Fig. 13.4. The time lapse between one generation and the next is about twenty minutes when the conditions are good – that is, when the bacteria have moisture, food and warmth. The division of bacterial cells give rise to a *colony*. Look at Fig. 13.5.

If conditions are not favourable for growth, some bacterial cells can change into resting *spores*. A tough coat forms around the nucleus, the cell material breaks down and only a small resistant spore remains. A bacterium may lie dormant as a spore for many years, and can resist difficult conditions like heat, cold or dryness. When conditions become favourable once more, the spore germinates to produce a vegetative cell and growth proceeds. Look at Fig. 13.6.

Like fungi, some bacteria live on dead organic material and others on living organisms.

Let's consider a few terms which are often used when referring to microbes:

- a microbe is *pathogenic* if it causes a disease
- a microbe is *commensal* if it lives in a living organism and the organism benefits from the relationship.

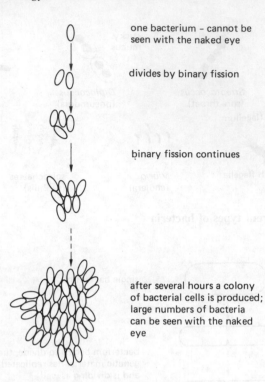

one bacterium – cannot be
seen with the naked eye

divides by binary fission

binary fission continues

after several hours a colony
of bacterial cells is produced;
large numbers of bacteria
can be seen with the naked
eye

Figure 13.5   Formation of a bacterial colony

⚠️

EXPERIMENT

Hypothesis to be tested: There are fewer bacteria on hands that have been
washed

**Work with micro-organisms can be very dangerous. Sometimes you cannot be sure
what you are growing! Never touch cultures you have grown, or expose them to the
air. They should be disposed of safely.**

You will need:
3 sterile Petri dishes
sterile nutrient agar in a screw-top
bottle kept in a warm water bath
bunsen burner
labelling pen
towel

disinfectant
cotton wool
plastic bag for waste
soap
access to a sink

Preliminary notes:
Sterilisation results in the complete absence of microbes. There are two methods:

- using an *autoclave*. This is like a large pressure cooker. The material is boiled
  under pressure. This destroys micro-organisms.
- using *radiation*. The materials are subjected to ultra-violet rays.

dormant spore

spore surrounded by
moisture, conditions
warm

spore takes in water
and swells

spore coat splits
and one vegetative
cell emerges

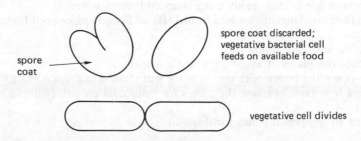

spore coat discarded;
vegetative bacterial cell
feeds on available food

spore
coat

vegetative cell divides

Figure 13.6   Germination of a spore

Method

1 Wipe down the work bench with cotton wool and disinfectant. Dispose of the cotton wool in the plastic bag.
2 Connect the bunsen burner and assemble the rest of the apparatus.
3 Label the Petri dishes on the bases as shown in Fig. 13.7.
4 Turn the Petri dishes up the 'right way'. Loosen the adhesive tape on one side of the dishes A and B. Do not open the dishes.
5 Remove the top of the bottle containing the agar. Pass the mouth of the bottle through the bunsen flame. This will kill any microbes on the lip of the glass.
6 Open the Petri dish A slightly. Make sure the opening is directed away from you. Quickly pour the agar into the plate to a depth of 0.5 cm. Close the dish and reseal.
7 Pour the agar into dish B in the same way.

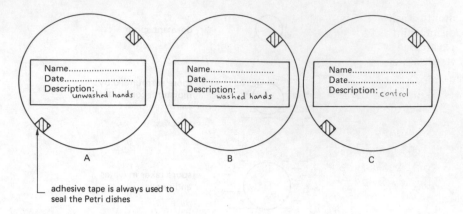

Figure 13.7   Labelling Petri dishes for microbiology work

8 Pass the mouth of the bottle through the flame once more and put the top back on it.
9 Leave the agar to set for a few minutes.
10 Open the plate labelled 'unwashed hands' (A) and gently place your fingers on to the agar.
11 Close the dish and reseal.
12 Wash your hands thoroughly using soap and warm water.
13 Open the plate labelled 'washed hands' (B) and gently place your fingers on to the agar.
14 Close the dish and reseal.
15 Incubate the plates at about 25 °C for a few days.
16 Wipe down the bench with cotton wool and disinfectant once more.
17 After a few days, examine the plates for microbial growth (without opening them).
18 Return all unopened plates for disposal.

Questions
1 Why is the bench washed down with disinfectant?
2 Why are the plates directed away from you when they are opened up?
3 Why are the plates resealed quickly after you have inoculated them?
4 What is the purpose of Petri dish C?
5 Do your findings verify your hypothesis?

EXPERIMENT

Hypothesis to be tested: Mouth wash prevents the growth of bacteria

You will need:
2 sterile Petri dishes
sterile nutrient agar in a screw-top
bottle in a water bath
2 pieces sterile filter paper

2 pair sterile forceps
bunsen burner
cotton wool
2 sterile pasteur pipettes

mouth wash
disinfectant
plastic bag
sterile water

wire loop
labelling pen
a culture of harmless bacteria
(e.g. *Bacillus subtilis*)

Method

1 Wipe down the bench with disinfectant.
2 Label the bases of the dishes:

    A – Mouth wash + *B. subtilis*   B – *B. subtilis*

  Also add your name and the date.
3 Pour agar into both plates as in the previous experiment.
4 Hold the wire loop in a hot bunsen flame until it glows red. Let it cool. Then use the loop to pick up a little *B. subtilis* from the culture plate. Gently stroke the loop to and fro over the agar in both the prepared plates (this is called *inoculating* the plates).
5 Pick up a piece of filter paper using a pair of sterile forceps. Pipette mouth wash over the paper and lay it on the agar in the centre of the correct plate.
6 Pipette sterile water over the second piece of filter paper and lay this on the agar in the second plate.
7 Close the plates and incubate at 25 °C for a few days.
8 Wipe down the bench with disinfectant.
9 After a few days, examine the plates for bacterial growth (without opening them).

Questions

1 Has the mouth wash stopped bacterial growth?
2 What was the purpose of the plate with the filter paper and water?

Design problem

Aim: To find the best conditions for microbial growth

You will need:
sterile Petri dishes
sterile water
disinfectant
plastic bag
labelling pen

dry bread
sterile pasteur pipettes
cotton wool
access to oven and refrigerator

Given the apparatus above, design an experiment to find out the best conditions for microbial growth. Check your plans with your teacher before beginning any practical work.

Summary

■ Microbes can only be seen with a microscope.

■ *Viruses* are small, parasitic organisms.

■ *Fungi* are either yeasts or filamentous.

- *Bacteria* have various forms. They reproduce by binary fission. Some bacteria can exist either as spores or as vegetative forms.

- Bacteria and fungi require *food*, *warmth*, *moisture* and *time* to grow.

Questions on Section 13.1

1 Explain how you would find out whether the surfaces in a kitchen were free from bacteria.
2 Look at Fig. 13.8.

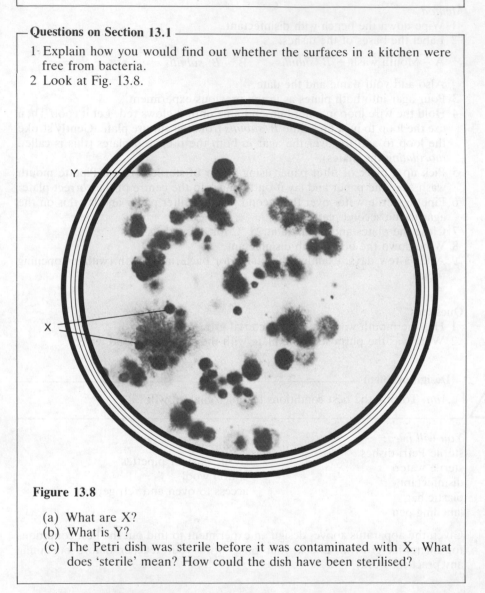

Figure 13.8

(a) What are X?
(b) What is Y?
(c) The Petri dish was sterile before it was contaminated with X. What does 'sterile' mean? How could the dish have been sterilised?

13.2 Spreading microbes around

Microbes are in the air and on most of the surfaces around us. Human beings and other animals play an important role in spreading microbes around and moving them from one place to another.

To cause a disease a pathogenic microbe must come into contact with the skin or the mucous membrances of the body. So how do microbes get to these surfaces? Look at Fig. 13.9. Notice how easy it is for microbes to be spread by human beings.

Table 13.1 lists the ways in which diseases are transmitted. Find these methods of disease transmission in Fig. 13.9 and take note of the diseases spread in each of these ways.

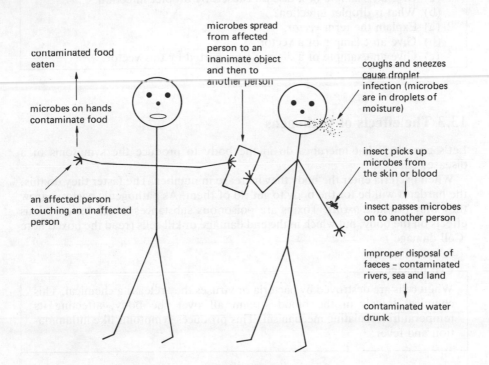

Figure 13.9   Human beings and the spread of microbes

Table 13.1   Some methods of transmitting diseases

Method	Description of method	Example
droplet infection	microbes are in droplets of moisture produced while talking, coughing and sneezing	chicken pox, colds, influenza, measles
contagion	microbes are spread by direct contact with a person or object carrying the microbe	impetigo, ringworm, colds
vector transmission	microbes are spread by animals biting and passing them into the body	malaria, rabies, plague

┌─ **Summary** ───┐
■ Microbes are spread by droplet infection, contact, vectors and contaminated food and water.
└──┘

┌─ **Questions on Section 13.2** ───────────────────────────────┐
1 (a) Give an example of a disease caused by droplet infection.
  (b) What is droplet infection?
2 (a) Explain the term *vector*.
  (b) Give an example of a vector.
  (c) Give an example of a disease transmitted by this vector.
└──┘

13.3 **The effects of pathogens**

Let's consider what microbes do in the body to produce the symptoms of a disease.

When bacteria enter the body they increase in number. The faster they do this, the harder it will be for the body to get rid of them. As pathogenic bacteria grow they may produce *toxins*. Toxins are poisonous substances which have various effects on the body, but which in the end damage or kill cells (read the boxed note 'Cell damage').

┌─ **Cell damage** ───┐
When cells are destroyed by bacteria or viruses they release a chemical. This chemical travels in the blood stream all over the body, affecting its temperature-regulating mechanism. This produces symptoms like inflammation and fever.
└──┘

Viruses invade cells of the body and destroy them. Look at Fig. 13.10. Notice that:

● the virus takes over the DNA of the cell and uses it to make more virus particles
● reproduction of viruses results in the destruction of body cells.

The symptoms of the disease are caused by the destruction of cells.

Fungi attack the skin or mucous membranes of the body. As they grow they cause irritation or inflammation, as well as other symptoms depending on the type of fungus – for example, breaking hair or flaking skin.

┌─ **Summary** ───┐
■ *Bacteria* release toxins which cause cell damage and death.

■ *Viruses* cause the destruction of cells.

■ *Fungi* attack cells in the skin or mucous membranes, which causes inflammation and irritation.
└──┘

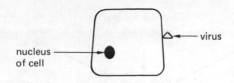

(a) Cell is invaded by a virus particle

(b) The virus takes over the DNA in the nucleus
and uses it to make more virus particles

(c) The newly formed virus particles are released from the cell,
and can then go on to invade more cells

Figure 13.10   Invasion of cells by viruses

Questions on Section 13.3

1 (a) What are toxins?
  (b) What organisms produce them?
  (c) What are the effects of toxins on the body?
2 Explain how viruses cause the symptoms of disease.

13.4 Food poisoning – cause and prevention

Microbes are on the food we eat. If there are not many of them or if they are of a harmless species, this is not a problem. But there are a few microbes that will cause food poisoning, especially if large numbers are present. We are going to consider two types of food poisoning bacteria: *Salmonella* and *Staphylococcus*.

Salmonella

This microbe is usually found in or on meat or fish products. Look at Fig. 13.11. Notice how many ways there are of contaminating meat with this bacterium. Fortunately *Salmonella* does not form spores and is killed in the normal cooking process. Cooking must therefore be thorough!

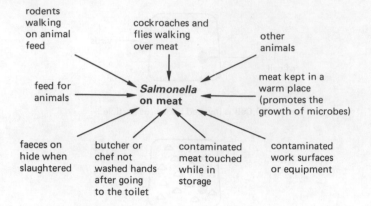

rodents
walking
on animal
feed

cockroaches and
flies walking
over meat

other
animals

feed for
animals

**Salmonella
on meat**

meat kept in a
warm place
(promotes the
growth of microbes)

faeces on
hide when
slaughtered

butcher or
chef not
washed hands
after going
to the toilet

contaminated
meat touched
while in
storage

contaminated
work surfaces
or equipment

Figure 13.11   Contamination of food with *Salmonella*

Think about a frozen chicken. Most chickens contain some *Salmonella*. If the chicken is put in the oven to roast before it is thoroughly thawed out the temperature in the centre of the bird may never get high enough to kill bacteria. In fact it may be just right for microbial growth – even though the outside of the chicken may look cooked.

Exercise

1 Your friend has never cooked a frozen chicken before. What advice would you give him?
2 Look at Fig. 13.12 which represents the arrangement of food in a refrigerator. Do you think the cooked meat should be on the bottom shelf as in A, or the top shelf as in B?

raw
meat   trifle

cooked
meat

cooked
meat   trifle

raw
meat

A   B

Figure 13.12   Storage of food in a refrigerator

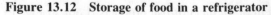

Hint: Remember that raw meat is often a good source of *Salmonella*.

Contamination of one food by another, as in this exercise, is called *cross-contamination*. It does not only occur in refrigerators.

Exercise

1 Study Fig. 13.13. Write down how the food eaten at 1 p.m. could carry microbes.

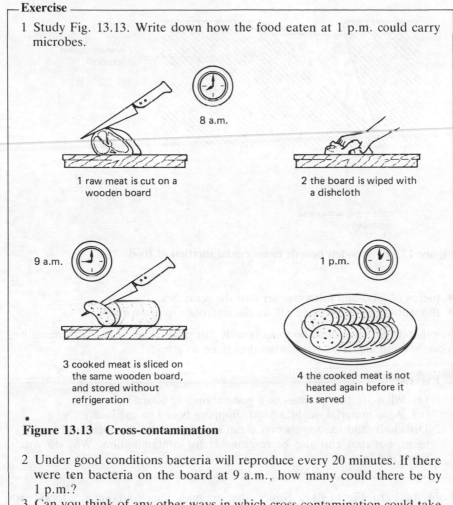

8 a.m.

1 raw meat is cut on a wooden board

2 the board is wiped with a dishcloth

9 a.m.

1 p.m.

3 cooked meat is sliced on the same wooden board, and stored without refrigeration

4 the cooked meat is not heated again before it is served

Figure 13.13 Cross-contamination

2 Under good conditions bacteria will reproduce every 20 minutes. If there were ten bacteria on the board at 9 a.m., how many could there be by 1 p.m.?
3 Can you think of any other ways in which cross-contamination could take place?

If cooked foods are contaminated by microbes they carry a high risk of causing food poisoning because they probably will not be cooked again. If the cooked foods are reheated the temperature may not be high enough to kill microbes. Look at Fig. 13.13 again. Notice that the chopping board was made of wood. Microbes can grow easily on wooden chopping boards. Look at Fig. 13.14. Notice that

● wood is porous
● wood is easily scratched, making crevices for microbes

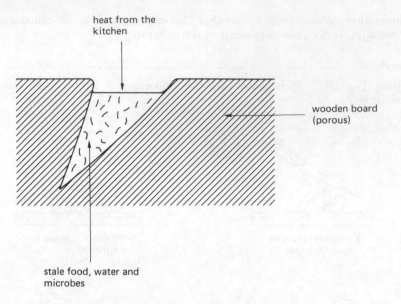

heat from the
kitchen

wooden board
(porous)

stale food, water and
microbes

Figure 13.14   Wooden boards cause contamination of food

- pieces of food and microbes get into the scratches
- the warmth of the kitchen allows the microbes to grow quickly.

In other words, wooden chopping boards can provide microbes with moisture, food, warmth and time – just what they need to grow!

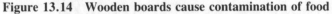

Exercise

1 (a) What are the qualities of a good chopping board?
  (b) What material might a good chopping board be made of?
2 Dishcloths and tea-towels may clean and dry kitchen surfaces and equipment, but they can also be responsible for contamination. Why do you think this is so?

Look back to Fig. 13.11. Notice that animals can contaminate food with *Salmonella* (and other microbes). Hairs on an animal's body increase its surface area, and allow it to carry more microbes.

Exercise

1 Make a list of all the hairy animals you can think of that may contaminate food in the kitchen.
2 Consider each animal on your list. Where might these animals have picked up pathogenic organisms?

While you may welcome your dog or cat into your kitchen, you would not welcome a cockroach, a mouse or a fly. But these unwelcome animals will only inhabit kitchens and houses where they can find food – for example, where crumbs are left lying around. Remember, they do not only contaminate the food they eat, but the equipment they walk on as well.

Flies deserve special mention. They get their food from excrement, decaying organic material like dead animals and plants and our food. Flies contaminate food in several ways:

- they walk over excrement and then over food and kitchen surfaces; their hairy bodies pick up lots of microbes
- they excrete on food and kitchen surfaces
- they eat by pouring saliva on to their food. The saliva digests it, and the fly then sucks it up. Look at Fig. 13.15. Notice that the tube (proboscis) which pours saliva over the food is the same as the one which takes up the digested products. If the fly has just sucked up digested excrement, microbes will stick to the tube. If the fly then lands on your food, flooding it with saliva, the microbes will be flushed out on to your meal.

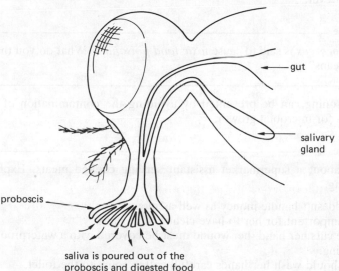

proboscis

gut

salivary gland

saliva is poured out of the proboscis and digested food is sucked up into it

Figure 13.15   How flies contaminate food

If food contaminated with *Salmonella* is eaten, symptoms of food poisoning may appear from one to 36 hours afterwards. Symptoms will only develop if large numbers of the microbe have been eaten, as only then can many bacteria survive the acidity of the stomach and reach the near-neutral conditions of the small intestine. In this environment *Salmonella* will reproduce. When they die they release poisonous toxins which cause fever, headache, diarrhoea and vomiting. *Salmonella* causes *infective food poisoning*, that is, the bacterium has to grow in the body before it causes symptoms of poisoning.

Staphylococcus

Staphylococcus lives around wounds, burns, ulcers and other damp warm places in the body. It is very easy, therefore, to transfer it from one's own body to food.

Exercise

Why do bacteria thrive around cuts and grazes?

Food can be contaminated with this bacterium if the cook touches his nose, mouth, ulcers or wounds during food preparation. Smoking while cooking is undesirable because the smoker's hands make contact with cigarette ends which have been in the mouth.

   Once the food is contaminated with *Staphylococcus*, the bacterium reproduces. It releases a toxin which causes vomiting from 8 to 12 hours after eating the food. *Staphylococcus* does not form a spore and is killed by heat: one or two minutes in boiling water is sufficient to kill it. The toxin can withstand 30 minutes in boiling water, however.

Exercise

Staphylococcus is said to cause *toxic food poisoning*. What do you think this term means?

Food poisoning can be prevented by avoiding the contamination of food and conditions for microbial growth.

Exercise

Think about a supermarket assistant serving cooked meats. Explain the following.

1 She doesn't handle money as well as meat.
2 It is important for her to have clean, short nails.
3 If she cuts her hand, her wound must be covered with a waterproof (blue) dressing.
4 She should wash her hands carefully after going to the toilet.
5 Her hair should be covered and tied back.
6 Her overall should be spotless.
7 She shouldn't smoke between customers.
8 She shouldn't lick her fingers.

Summary

■ *Salmonella* is a food-poisoning bacterium usually found on or in moist protein foods like meat and eggs.

■ Frozen chickens must be completely thawed before cooking.

■ The transfer of *Salmonella* from raw food to cooked food is called *cross-contamination*.

- *Salmonella* causes infective food poisoning.
- *Staphylococcus* is a food-poisoning bacterium found in the mouth and nose and around burns and wounds.
- *Staphylococcus* can be transferred to food by a person touching their nose, for example, and then touching food.
- *Staphylococcus* causes toxic food poisoning.

Questions on Section 13.4

1 The temperature at which meat pies are kept warm in a cafeteria, is very important. Why must the temperature be high?
2 I took a chicken leg from the freezer this morning. This evening I decided I didn't need it so I put it back into the freezer for use another day. Why was this a risky thing to do?

13.5 **Food preservation**

Food preservation is practised so that

- food is safe for consumption for a long time
- food can be exported
- food is available out of season
- food may be given different flavours.

The methods of preservation are based on the knowledge of physical and chemical factors influencing the growth of micro-organisms. There are five categories of food preservation methods:

- high-temperature preservation
- low-temperature preservation
- dehydration (drying)
- chemical preservation
- irradiation.

Let's consider high-temperature preservation first. High temperatures denature enzymes and kill microbes.

Exercise

1 Look at Fig. 13.16. At what temperature are yeast and mould spores destroyed?
2 At what temperature are vegetative bacteria destroyed?
3 What is the minimum temperature necessary to kill bacterial spores?

Pasteurisation is a short-term method of food preservation. Foods like milk and cream are heated to temperatures that will kill the vegetative forms of microbes. Clearly, after a time any spores present will germinate and the food will then carry

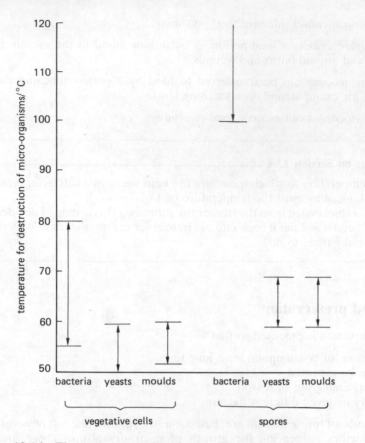

Figure 13.16   The temperature required to kill micro-organisms (moist heat)

increasing numbers of vegetative forms. This is why pasteurised milk will go sour. Additional methods of preservation like refrigeration are used to keep pasteurised products safe for a little longer. There are two main methods of pasteurisation:

- HTST (high temperature, short time): the food is heated to 71.7 °C for 15 seconds
- LTH (low-temperature holding): the food is heated to 66.8 °C for 30 minutes.

Sterilisation techniques raise food temperatures over 100 °C, usually using steam under pressure. This completely destroys micro-organisms.

Milk can be sterilised, but the procedure alters its flavour and appearance. UHT milk is 'ultra heat treated', that is, it is heated to 135 °C for two seconds. This kills all vegetative forms and spores. Canning food also uses heat treatment.

Low-temperature preservation works in quite a different way. Low temperatures do not kill microbes, but they slow down their metabolism. The lower the temperature falls, the less will be the microbial activity. There are various degrees of low-temperature preservation:

- cellar storage – suitable for short-term storage of foods such as vegetables

- chilling – refrigeration between 1 and 4 °C, slowing down the growth of pathogens and spoilage organisms
- freezing – uses temperatures of −18 °C to −23 °C. These temperatures stop the multiplication of microbes. They may destroy some more delicate vegetative forms, but all spores can survive.

Dehydration reduces the amount of water in the food to a level below which micro-organisms cannot grow and multiply. Once water is added to the food, microbes will start to grow again.

Chemical preservation prevents microbes from growing on food because

- it interferes with cell permeability
- it interferes with the cell's enzyme activity
- it interferes with the genetic mechanism of the cell.

Sugar and salt are traditional chemical preservatives which work similarly. They dehydrate the microbes by osmosis, thus preventing further activity.

■ **Memory check**
What is osmosis?
See Section 2.1.

Lactic acid is formed in the fermentation of cabbage to produce sauerkraut. It is also produced by bacteria in yoghurt formation. Acids act as preservatives because enzymes are sensitive to pH and microbial enzymes are therefore affected by acidic conditions. Vinegar (ethanoic acid) also works in this way.

Alcohol is another traditionally used preservative. Smoked food is also unsuitable for microbial growth. Some chemicals have only been used as preservatives quite recently. These include sulphur dioxide in sausages, benzoic acid in fruit juices, and chlorine in water purification.

Antibiotics such as nisin are sometimes used to preserve cheese and cream. Tetracyclines can be used in small amounts to preserve fish, but they are not permitted for use with poultry and meat carcasses in the UK.

Radiation (cold sterilisation) can be used to preserve food because the rays damage the nuclei of microbes, preventing cell division. Ionising radiations such as X-rays, beta-rays and gamma-rays are used. These also slow down the ripening of fruit and vegetables and the sprouting of potatoes, and kill insects infesting grain. Ultra-violet radiation causes mutation and death in microbes. These rays cannot penetrate opaque material, however, so the effect is only on the outside of the irradiated product. Ultra-violet rays are used to purify air in rooms, and to kill spores on crystal sugars and packaged cheese and yeasts growing on pickles.

┌─ **Design exercise** ─────────────────────────────
You are provided with the following equipment:

vinegar	2 beakers
distilled water	2 mushrooms

Design an experiment to find out if vinegar can be used to preserve mushrooms. Either state the aim of your experiment, or construct a hypothesis to test.

┌───┐
│ ┌─ **EXPERIMENT** ──│
│   │
│ *Hypothesis to be tested*: Bacterial numbers in milk increase with the age of │
│ the milk   │
└───┘

This experiment uses a dye called *resazurin*, which is used in dairies to find out if their milk is fresh. The dye is blue in fresh milk that contains few bacteria. As bacteria grow they use up the oxygen in the milk and the dye changes to a pink colour and finally becomes colourless. The quicker the dye changes colour the more bacteria must be present.

You will need:
1 cm³ syringe   5 10 cm³ syringes
10 cm³ resazurin dye   5 samples of pasteurised milk of
5 test tubes   different ages
test tube rack

Method
1 Label the test tubes A, B, C, D and E.
2 Put 10 cm³ milk into each tube. Use a different syringe for each milk sample. Record the age of the milk put into each tube.
3 Place 1 cm³ resazurin dye into each milk sample. Swirl the contents.
4 Observe and record the colour of each mixture immediately.
5 Allow the milk samples to stand for 15 minutes and then record the final colour of each mixture. Construct a results table.

Questions
1 Does bacterial contamination increase with the age of the milk?
2 Why is it important not to shake the tubes after the initial mixing?
3 It is thought that as milk sours bacteria change lactose into lactic acid. Spoilage of milk could be indicated by a change in pH. Write a hypothesis and design an experiment to find out if pH does change significantly as milk sours. List all the equipment you need, how you are going to do the experiment and what you are going to record. Include a results table.

┌─ **Summary** ──┐
│   │
│ ■ There are two types of *high-temperature preservation*. *Pasteurisation* kills vegetative forms of microbes. *Sterilisation* kills both vegetative and spore forms. │
│   │
│ ■ *Low-temperature preservation* may slow down or stop microbial growth. │
│   │
│ ■ *Dehydration* preserves food because microbes cannot grow without water. │
│   │
│ ■ *Chemicals* like sugar, salt, vinegar, benzoic acid and sulphur dioxide prevent microbial growth. │
│   │
│ ■ *Radiation* prevents microbes from reproducing.   │
└──┘

13.6 **We wouldn't be without them**

We often think of microbes as nasty germs that make our food bad and make us
ill. But we need the microbes too. For example:

- micro-organisms living in the large intestine provide us with vitamin K
- micro-organisms decompose plant and animal matter and release nitrates and
  carbon dioxide which are important for plant growth (see Section 4.1); not only
  does this get rid of the dead organic material – it also enables new plants and
  animals (including ourselves) to grow
- micro-organisms are used by human beings to make many useful products.

Let's consider some of the products of micro-organisms in more detail.

Fungi have been used to make wine and beer for centuries. Yeasts are added to
a sugary or starchy material and alcohol is formed. The process is called
fermentation. It can be summarised by the equation

$$\text{glucose} \longrightarrow \text{alcohol} + \text{carbon dioxide}$$

Fermentation by yeast also takes place in bread dough. The carbon dioxide
produced by the yeast causes the bread to rise – any alcohol produced evaporates
during cooking.

■ **Memory check**
'Fermentation' is a term describing respiration by this yeast. What can you
remember about anaerobic and aerobic respiration?
See Section 7.5.

Fungi are also used in the production of *antibiotics*. These are medicines that
kill bacteria. The most common antibiotic is *penicillin*, which is produced by the
fungus *Penicillium*. The fungus is grown in huge tanks, each containing 250 000
litres of nutritive fluid. Temperature and pH are carefully controlled so that the
best growing conditions are maintained. The antibiotic is produced during a
certain stage in the fungus life cycle. It is separated from mycelium and unused
solid nutrients, dried and crystallised or dissolved in a suitable liquid. After

purification it is made into tablets or liquid medicines. Other antibiotics made by fungi include streptomycin, tetracycline and aureomycin.

Bacteria are used to make cheese. In cheesemaking, a bacterial 'starter' is added to pasteurised milk to sour it. The type of starter used gives the cheese its particular flavour. Rennet is added to the soured milk, causing solid curds and liquid whey to form. (Rennet is an enzyme preparation made either from the stomach of young calves or artificially by microbes. You read about it in Section 2.2). The curds are separated from the whey. The solid curd is squeezed to remove more water and then it is left to ripen. Fungi are encouraged to grow on some cheeses; these form blue-green veins in the cheese, or a skin of fungi. The fungi also give particular flavours to the cheeses.

Bacteria also convert beer into vinegar. Young beer trickles continuously down a tower packed with wood shavings. Bacteria growing on the shavings turn the beer into vinegar by changing alcohol into ethanoic acid.

Bacteria are used to make yoghurt. They convert the milk sugar lactose to lactic acid. The lactic acid sours and thickens the milk.

Using microbes to produce foods or drugs is a part of *biotechnology*.

EXPERIMENT

Aim: To find out which sort of milk can be used to make yoghurt

---

You will need:

3 100 cm³ sterile beakers	100 cm³ UHT milk
1 large sterile spatula	100 cm³ pasteurised milk
3 sterile glass rods	100 cm³ reconstituted dried milk (dried
clingfilm	milk dissolved in water, according to the
water bath set at 25 °C	package directions)
small carton natural yoghurt	Universal indicator paper

Method
1 Label the beakers A, B and C.
2 Pour the UHT milk into beaker A, the pasteurised milk into beaker B and the reconstituted dried milk into beaker C.
3 Find the pH of the milk in each beaker, using Universal indicator paper.
4 Add a spatula of yoghurt to each beaker.
5 Mix the yoghurt into the milk with the glass rods. Use a different glass rod for each beaker. Put clingfilm over each beaker.
6 Incubate the beakers in a water bath overnight. Construct a results table.
7 Examine each beaker and record your observations. Determine the pH of the mixture in each beaker.

Warning: Do not taste yoghurt made in a laboratory.

Questions
1 Which type of milk was best for making yoghurt? Explain your answer.
2 Design experiments to test the following hypotheses:
   (a) The amount of yoghurt added to the milk does not affect the quality of yoghurt produced.
   (b) The type of yoghurt used as a starter is important to the quality of the product.

┌───┐
EXPERIMENT

Aim: To find out which antibiotic is most effective in killing *Bacillus subtilis*
└───┘

You will need:

2 nutrient agar plates	bunsen burner
a wire loop	cotton wool
culture of *Bacillus subtilis*	disinfectant
discs impregnated with antibiotics	waste disposal bag
sterile forceps	

Method

1 Wipe down the work surfaces with disinfectant and cotton wool.
2 Inoculate both plates with the *Bacillus subtilis*. (Refer to Section 13.1 if you have forgotten how to do this.)
3 Using the sterile forceps place the impregnated discs on to the surface of the agar.
4 Seal the plates. Make sure they are labelled. Incubate for a few days.
5 Construct a results table. Think carefully how about what you are going to record.
6 Return all unopened plates for disposal.

Questions

1 Which antibiotic killed *Bacillus subtilis* the best?
2 Why is it useful to know that more than one antibiotic may be effective in killing a particular type of bacterium?

┌───┐
Summary

■ Fermentation can be summarised as:

glucose ⟶ alcohol + carbon dioxide

■ Fermentation is used in the production of wine, beer and bread.

■ Some fungi are grown industrially to produce antibiotics, such as penicillin, streptomycin, aureomycin and tetracycline.

■ Bacteria are used in the production of cheese, vinegar and yoghurt.
└───┘

13.7 Defence and attack

Despite all the good things microbes can do for us, there is no escaping the fact that some of them do cause disease. The first question to ask is: how do they get into our body? Look at Fig. 13.17. Notice where microbes can get into the body and how the body is well designed to keep them out and kill them should they get in.

Despite these defence mechanisms, sometimes microbes get into the body and cause disease. Recovery depends on destroying the microbes invading the body. As you read in Section 6.1, the body produces *antibodies* to do this. Let's remind ourselves that antibodies are:

first line of defence

respiratory system   digestive system   skin

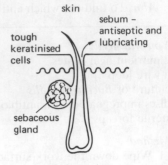

sebum –
antiseptic and
lubricating

tough
keratinised
cells

sebaceous
gland

ciliated   mucus-producing
cell   cell   digestive glands

microbes become   acid from the   skin is tough,
stuck to mucus;   stomach and   waterproof and
cilia move mucus   digestive   antiseptic
and microbes to   enzymes kill
exterior   microbes

second line of defence

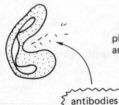

phagocytes engulf microbes,
antibodies make them harmless

antibodies

Figure 13.17   First and second lines of defence

- proteins
- produced by the lymphocytes
- released into the plasma
- *specific*: that is, one antibody is made for one type of microbe – for example, the chicken pox antibody is ineffective against measles or any other infectious disease; it is only effective against chicken pox
- produced in response to a microbe or foreign substance in the body; substances stimulating antibody production are called *antigens*.

■ **Memory check**
How do antibodies affect microbes?
See Section 6.1.

Once the body has learnt how to recognise a particular antigen and make antibodies against it, it will not forget. The body is said to be *immune* to that antigen.

There are some diseases, however, which are either fatal or likely to leave us with varying disabilities – if, indeed, we should survive. For example, small-pox – though now extinct – was a particularly distressing and disfiguring disease which was very often fatal. We can get protection against many such diseases by *vaccination*.

A vaccine is made from micro-organisms that have been killed or weakened by treatment with, for example, heat, formaldehyde or ultra-sound. These 'attenua-ted' organisms are harmless, but the body will still react to them and make the same antibodies that it would make during an actual attack of the disease.

Babies and young children are vaccinated against polio, diphtheria, tetanus, whooping cough, measles, mumps and rubella.

Exercise

Rubella (German measles) is a disease caused by a virus that can pass across the placental wall and into a developing foetus. In the early months of pregnancy contact with rubella can cause brain damage, blindness or deafness in the foetus. Up until 1989 only adolescent girls were offered the vaccine.

1 Why do you think the vaccine was offered to girls only?
2 Why is the vaccine now given to both boys and girls?

Note that the body acquires immunity to an antigen when it is able to recognise it and make antibodies against it. Such immunity can be acquired in two ways:

● getting the disease and surviving it
● vaccination.

If a person has been in contact with a dangerous disease, such as rabies, there is not time for the body to make antibodies. In this case 'ready-made' antibodies are given by injection. These antibodies should make the antigen harmless and the person may well survive. As the body has not learnt how to make the antibodies, however, the immunity only lasts as long as the injected antibodies are circulating in the body. There are thus two types of immunity:

● *active immunity* – where the body makes antibodies
● *passive immunity* – the body is given antibodies but does not learn how to make them.

Look at Fig. 13.18. Notice that the antibody levels are high immediately after the injection but after a while the person loses the immunity.

Look at Fig. 13.19. Vaccinations against diseases like polio, diphtheria and tetanus are given as a course – that is, several doses of vaccine are given at prescribed intervals. Notice from Fig. 13.19 that:

● the body responds to a second injection of a vaccine much more quickly and with greater effect than to the first dose
● the person is left with a degree of immunity.

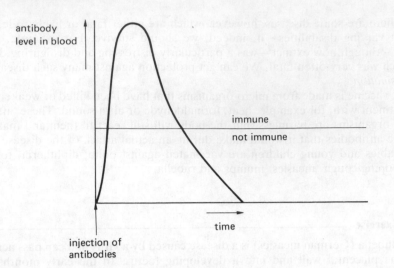

Figure 13.18   Development and loss of passive immunity

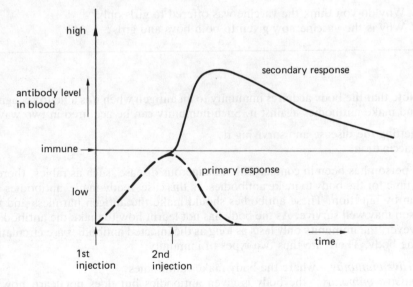

Figure 13.19   Active immunity through two doses of vaccine

Exercise

1 Babies are not vaccinated until they are three months old. Are they protected against disease during the first three months of life? If you said 'yes' to this question, explain how you think they are protected.

2 When a rhesus-negative woman gives birth to a rhesus-positive baby, she is given rhesus antibodies. Why is this? Is this active or passive immunity? (Read Section 6.6 again if necessary.)

Microbes can be killed outside the body by:

- heat – usually moist heat is used to sterilise equipment, that is, all microbes are destroyed; this method is used by some hairdressers and chiropodists and also in hospitals
- chemicals – microbes can be reduced to harmless numbers on equipment by using *disinfectants*; *antiseptics* are generally used on the skin
- radiation – ultra-violet light is used by some hairdressers, and ionising radiation is used to sterilise medical equipment.

Summary

■ The body resists the entry of microbes by mucus and ciliary action in the breathing passages and by the acid in the stomach, and also because the skin is tough, waterproof and antiseptic.

■ If microbes get into the body they are either engulfed by phagocytes or are made harmless by antibodies.

■ If the body is able to recognise a microbe and respond by making the appropriate antibody, it is said to be immune to that antigen. This is *active immunity*.

■ Recovery from a disease can be brought about by treatment with ready-made antibodies. In this case the body does not make its own antibodies, and cannot make antibodies if it comes into contact with the same microbe again. This is *passive immunity*.

Questions on Section 13.7

1 Explain the following terms:
   (a) antigen
   (b) antibody
   (c) passive immunity
   (d) active immunity.
2 In the early twentieth century the death rate from diphtheria was very high. Nowadays hardly anyone contracts diphtheria. Explain why this is so.

Methane, typhoid and rubbish

14.1 Sewage

Human beings produce sewage. Sewage is waste carried away by water. The constituents of sewage fall into three groups:

- domestic waste: faeces and urine
  bath water
  water from washing clothes
  kitchen water
- industrial waste: the type depends on the local industry, but it may contain heavy metals like mercury, zinc and lead
- surface water: rain water collected from roads.

Sewage contains substances which are harmful to humans and the environment. These substances are:

- pathogenic bacteria
- industrial chemicals like heavy metals
- chemicals in detergents
- organic matter.

Sometimes sewage is discharged straight into rivers, lakes or the sea. Look at Fig. 14.1. Notice the effect of large quantities of organic matter on freshwater waterways.

Exercise

Look at Fig. 14.1.

1 What effect does the sewage have on oxygen concentration in the stream?
2 Why does the sewage have this effect?
3 What effect will the changing concentration of oxygen have on animals like fish in the stream?

Fig. 14.1 does not even consider the effects of other harmful substances in sewage! Discharge of sewage into the sea results in unsightly beaches, unsafe swimming and infected shellfish that can cause food poisoning. Bacteria responsible for causing typhoid, cholera and dysentery may be present in sewage; so too may the viruses causing poliomyelitis and infectious hepatitis.

If sewage treatment is carried out its aims are:

- to reduce the *biological oxygen demand* (read the boxed note 'Biological oyxgen demand')
- kill the pathogenic micro-organisms.

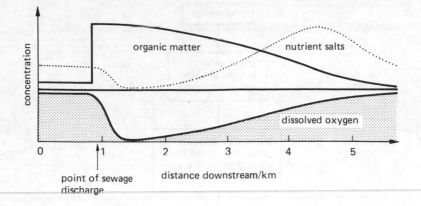

Figure 14.1 The effect of sewage on the oxygen concentration in a stream

Sewage treatment may not take out the industrial waste or any other harmful substances.

> **Biological oxygen demand**
>
> As organisms like animals, fungi and aerobic bacteria grow in ponds, lakes and streams they use up oxygen. In 'healthy' ponds, lakes and streams, the amount of oxygen produced by plants and dissolving into the water from the atmosphere is enough for the organisms living in it. But if nitrates and other organic matter are available in larger than 'normal' amounts, bacteria and fungi will grow too fast and use up the oxygen too quickly.
>
> The demand on oxygen by the organisms living in a body of water is called the biological oxygen demand (BOD for short) of that water.

Look at Fig. 14.2. This shows the preliminary treatment of sewage diagrammatically. Notice that the sewage is separated into two components: *sewage sludge* and *liquid sewage*. Take note of what these contain.

There are two different ways of treating liquid sewage. Let's consider the *trickling filter system* first. Look at Fig. 14.3. Notice that the trickling filter system encourages the growth of *aerobic* organisms. Aerobic organisms need oxygen, so spraying liquid sewage over the stones brings more oxygen into the filters. The organisms growing on the stones break down the sewage constituents as follows:

> carbohydrates $\longrightarrow$ carbon dioxide + water
> proteins $\longrightarrow$ ammonia + carbon dioxide + hydrogen sulphide
> lipids + soaps $\longrightarrow$ carbon dioxide + water
> urea $\longrightarrow$ ammonia + carbon dioxide
> ammonia $\longrightarrow$ nitrates + water
> hydrogen sulphide $\longrightarrow$ sulphates + water

Notice that, in general, the organisms break down large molecules into smaller molecules. Protozoa (one-celled animals) in the filter eat bacteria in the sewage liquid.

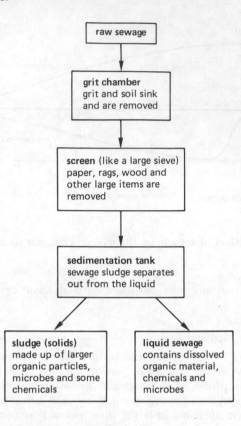

Figure 14.2   Preliminary treatment of sewage

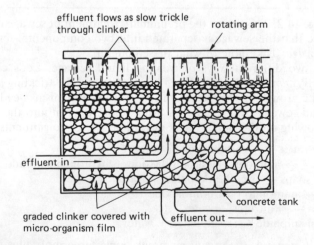

Figure 14.3   Trickling filter system of sewage treatment

The resulting water is quite harmless and can be discharged into the sea or a river.

Now let's consider the *activated sludge process* of sewage treatment. Look at Fig. 14.4.

Notice that in this process the liquid is aerated. As a result of this aeration, particles of organic matter called *floc* form. Floc is made up of bacteria and protozoa, and works in the same way as the microbial layer over the stones in the filter method. That is, it breaks down organic material and reduces the numbers of bacteria. The activated sludge method works more quickly than the filter method, however, and is used by about half the sewage treatment plants in the United Kingdom.

The sewage sludge is usually passed into a large tank where it can be digested by *anaerobic* micro-organisms. These organisms can grow without any oxygen. Look at Fig. 14.5. Notice that the methane produced by the micro-organisms is used to assist the digesting process – methane is a good fuel. Also take note of what happens to the digested sludge.

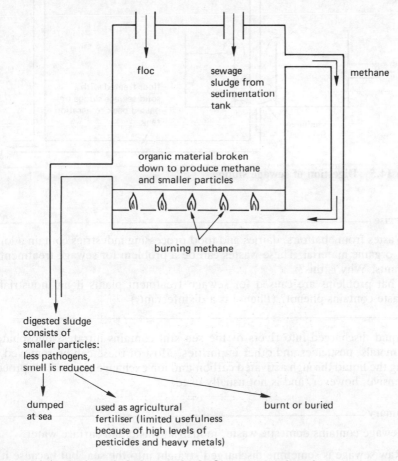

Figure 14.4   Activated sludge method of sewage treatment

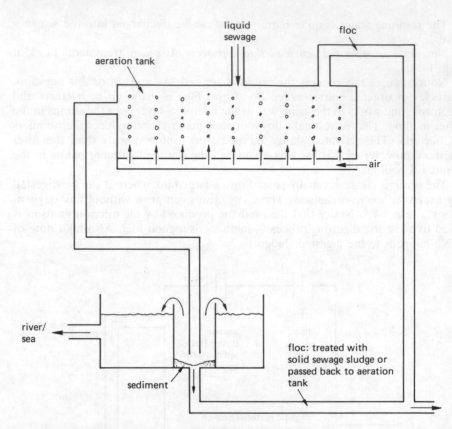

Figure 14.5   Digestion of sewage sludge

---Exercise---

1 Wastes from abattoirs, dairies and food processing industries contain a lot
  of organic material. These wastes can be a problem for sewage treatment
  plants. Why is this?
2 What problems are caused for sewage treatment plants if an industrial
  waste contains phenol? (Phenol is a disinfectant.)

The liquid discharged into rivers or the sea still contains nitrates, phosphates,
heavy metals, pesticides and other impurities. Most of these can be removed by
filtering the liquid through activated carbon and ion exchange resins. This process
is expensive, however, and is not usually used.

---Summary--

■ Sewage contains domestic waste, industrial waste and surface water.

■ Raw sewage is sometime discharged straight into the sea, but because it
  contains pathogenic micro-organisms it should be treated first.

- Sewage is
  allowed to settle to remove grit and soil
  screened to remove large items like wood, rags and paper
  allowed to settle again so that solid sludge separates from the liquid
  sewage.

- Liquid sewage is treated by either the trickling filter or the activated
  sludge method. Both processes rely on *aerobic* micro-organisms to break
  down organic material and protozoa to reduce the numbers of pathogenic
  organisms.

- Solid sludge is usually digested by *anaerobic* bacteria. Methane is
  produced and the sludge is broken down into smaller particles. The
  process also reduces the smell and the numbers of pathogens.

- The digested sludge is dumped at sea, buried or put on to agricultural land
  as fertiliser.

Questions on Section 14.1

1 Describe the roles of (a) aerobic and (b) anaerobic micro-organisms in the
  treatment of sewage.
2 (a) What is sewage?
  (b) What is the purpose of sewage treatment?
3 Why are some areas of Britain finding it difficult to treat sewage
  effectively?

PROJECT WORK

Visit your local sewage treatment works.

Find out what sort of treatment they use.

Find out if there are any particular problems that this plant has to cope with.

What do they do with their digested sewage sludge?

14.2 Water

Let's start this section by considering where our water comes from.

Exercise

In the form of a diagram, describe where you think your water comes from.

Compare your diagram with Fig. 14.6 on the next page.

From your own diagram and Fig. 14.6 you can see that there are three different
ways by which water can be collected:

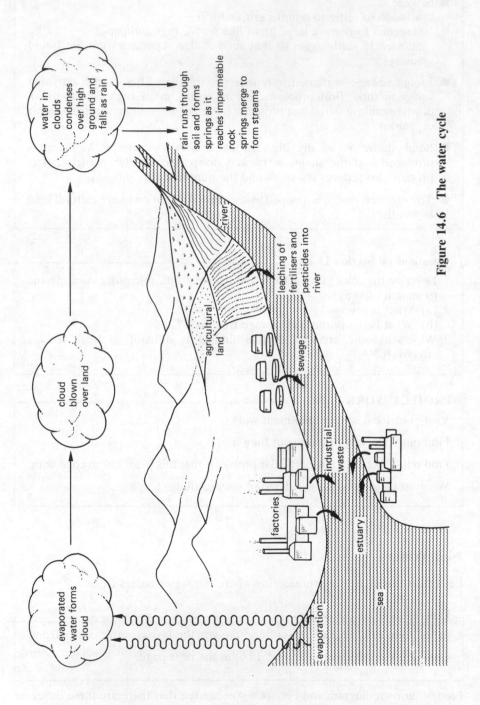

Figure 14.6   The water cycle

- rain
- water from reservoirs, lakes, rivers and streams
- ground water.

Exercise

Look at the three ways by which water can be collected.

Which type of water will be the cleanest? Why?
Which will be the most polluted? What will cause this pollution?

Read on to check your answers.

Rain water contains impurities from the atmosphere. In inner cities or in industrial areas, for example, it contains many impurities that make it unsuitable for drinking (see Section 4.4).

Water from streams, rivers and lakes varies in its purity. Water collected from high ground may be fairly free from bacteria and unpolluted. Water collected from lower ground may be contaminated with sewage (which may contain pathogenic bacteria), fertilisers washed from farmland, pesticides, industrial effluent and domestic detergents. Some of these will contain toxic substances. Water collected from low ground needs purification.

Ground water has soaked through the ground and is supported by an impervious layer of rock. Where this impervious layer is in contact with the surface of the ground – as on the slope of a hill – it emerges as a spring. Alternatively, it can be tapped by sinking wells. The purity of ground water depends on what land the water has soaked through.

Many diseases can be spread by water. Usually these are a result of faecal contamination. They include typhoid, paratyphoid, dysentery, poliomyelitis and cholera.

Water samples in water treatment plants are tested for bacterial content. That is, scientists take samples of water and put them on agar plates, which are then incubated. From the growths on the plates, the numbers of bacteria of intestinal origin can be estimated, as an indication of the degree of sewage contamination.

Water purification produces water that is fit to drink. This water must be free from microbes, solid contaminants, harmful chemicals, smell and colour. Purification involves:

- sedimentation
- filtration
- disinfection.

Look at Fig. 14.7. Notice that the filtration of water is very similar to the trickling filter method of sewage treatment. The micro-organisms involved are the same but there is less oxygen in the system because the liquid is not sprayed over the surface of the filter.

Water may be treated in several other ways to improve its quality. For example, fluorine may be added to reduce tooth decay (see Section 5.2). But water purification does not usually remove substances like nitrates, heavy metals like lead or mercury (from industrial wastes entering rivers) or pesticides.

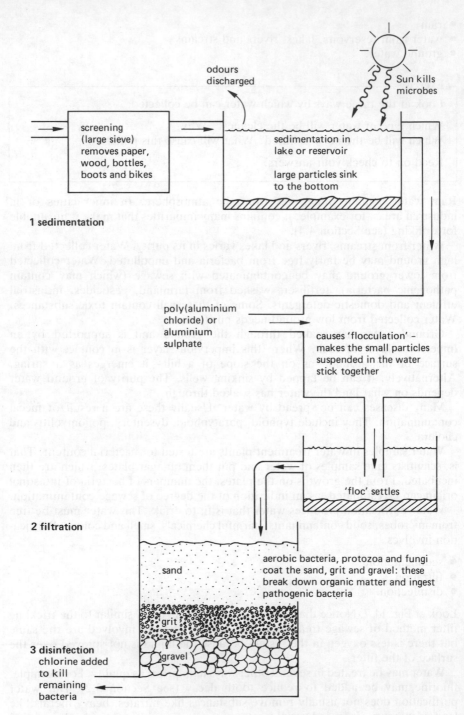

Figure 14.7   Treatment of water

Exercise

1 (a) Where do nitrates in drinking water come from?
  (b) Which areas of the country are likely to have more nitrates in their drinking water?
2 How does lead get into drinking water?

The effects of nitrates on the human body have not been fully established. Nitrates do however appear to be linked with 'blue-baby syndrome'. The guts of all baby animals, including humans, convert nitrates to nitrites. The nitrites combine with haemoglobin, preventing it from combining with oxygen. Oxygen carriage in the baby's body is thus reduced, to the point of asphyxiation in severe cases. This is why parents in some rural areas of England are advised not to give their babies tap water. There is also a suggestion that nitrates are linked with stomach cancer in adults.

If water containing lead is consumed, the lead is deposited in the bones – it is not eliminated. The levels of lead increase until they have toxic effects. In children lead interferes with the development of nervous tissue. Lead is more soluble in soft water than in hard water. The amount of lead coming from pipes in hard water areas is therefore less than in soft water areas.

Summary

■ Water comes from rain, rivers, reservoirs and the ground.

■ Diseases like cholera can be spread through water.

■ Water is purified to reduce the levels of bacterial contamination and make it fit to drink.

■ Water is purified in three stages: sedimentation, filtration and disinfection.

■ Filtration of water is similar to the trickling filter process in sewage treatment.

■ Nitrates and heavy metals have toxic effects if they are present in water in high enough concentrations.

Questions on Section 14.2

1 What is the purpose of the following stages in water purification:
  (a) sedimentation
  (b) filtration
  (c) disinfection?
2 What are the harmful effects of (a) nitrates, (b) lead in drinking water?
3 How do (a) nitrates and (b) lead get into drinking water?

┌─ **PROJECT WORK** ──────────────────────────────────────┐

Visit your local water purification plant.

Draw a plan of the plant.

Describe what happens to the water at each stage of the purification process.

What is the quality of the water they have to purify?

└───┘

14.3 **Rubbish, recycling and responsibility**

Rubbish consists of unwanted items that are to be disposed of.

Domestic refuse is a mixture of organic kitchen and garden waste and packaging, which may be plastic, glass or metal. There may be other components too.

Industrial refuse consists of packaging, organic waste, sawdust, metal scrap, poisonous materials (such as cyanide, mercury or phenol), radioactive waste and poor-quality rocks from mines and quarries.

The way in which refuse is disposed of varies in different parts of the country. There are four main methods of refuse disposal:

● incineration
● recycling
● landfill (burial)
● tight packing.

Incineration (burning) disposes of large amounts of refuse and provided ash or 'clinker' which can be used to make concrete. The fumes produces are toxic, however, and the process is not as popular as it was.

In some areas the refuse is thoroughly sorted. Where possible, the results of the sorting process are *recycled*. This means that they are used again. Recycling conserves the raw materials used to make these items. For example, recycling paper reduces the demand for wood and trees are saved. Combustible material that remains can be processed to make solid fuel pellets, which are then sold.

The controlled *tipping* of refuse, usually on low-lying waste land, remains the most widely used method of disposal. Rubbish is spread on the land in layers 1.5 m thick. Each layer is allowed to settle before another is added. The ground is sprayed with insecticide from time to time and the area surrounded by wire netting to keep out rats and dogs.

Tight packing of refuse by bulldozers prevents rats burrowing into it. It will still need spraying with insecticides to prevent infestation by flies as it slowly decomposes. When the tightly packed refuse reaches a certain height it is covered with soil.

Some people are making themselves more responsible for the rubbish they produce. They decide to sort certain items from their rubbish like glass, paper and aluminium (cans, ring pulls from drink cans and milk bottle tops). These items are then taken to bottle banks, paper depots and so on. In some areas Friends of the Earth and other organisations have depots where sorted rubbish can be taken. The materials can then be recycled.

Recycling can lower the cost of the item produced. When broken glass is reprocessed, the temperature required is less than that for new glass, so energy can be saved.

Some items, like milk bottles, are *re-used*. The cost of collecting, washing and sterilising a milk bottle is a fraction of the cost of making a milk bottle from scratch, even when recycled glass is used.

EXPERIMENT

Aim: To find out which components of rubbish decompose the quickest

1 Obtain small samples of different types of rubbish your household is throwing away.
2 Design an experiment to find out which of these samples decomposes the quickest.
3 Make a list of all the equipment you will need. Write down what you are going to do and construct a results table before you begin.
4 Check your plans with your teacher before beginning any practical work.

Summary

■ Rubbish can be incinerated
   sorted and recycled
   buried
   tightly packed.

■ Materials such as glass, paper and metal can be recycled, and in some cases re-used. Recycling and re-use reduce the demand for raw materials.

Questions on Section 14.3

1 Rubbish can be a health hazard. How does the disposal of rubbish reduce the likelihood of the spread of disease?
2 (a) What does 'recycling' mean?
   (b) What items can be recycled?
   (c) Why should recycling be encouraged?

PROJECT WORK

1 Find out what schemes exist in your area for recycling.
   Find out (a) what is recycled, (b) who does the recycling, (c) how the recycling takes place, and (d) what the items are recycled into.
2 Choose one item that is recycled in your area. Find out the cost of recycling a given amount of this item. Compare this to the cost of making the same amount of the item from the raw materials.
3 Find out how the rubbish in your area is disposed of.

⬡ Appendix

Hypotheses

The method by which all scientific 'knowledge' is built up is based on the writing and testing of hypotheses. Let's look at how this is done.

The first stage for a scientist is *to make observations*, and to record them carefully.

Imagine yourself in this situation. Read the following observations:

- The contents of a cell will break down carbohydrate like starch into smaller particles but will not break down protein.
- Saliva breaks down starch into smaller particles but will not break down protein.
- Secretions in the stomach will break down protein into smaller particles but will not break down starch.
- Secretions from the pancreas will break down starch and protein into smaller particles.

There are a lot of complicated observations here. They might be more useful if they were sorted or *classified*. If we sort them out we can present them in just two statements:

- The contents of a cell, saliva and pancreatic secretions all contain a substance that breaks down starch.
- The secretions from the stomach and pancreas both contain a substance that breaks down protein.

Notice that we have separated the breakdown of carbohydrate from the break-down of protein. This not only makes the observations easier to understand but also isolates two areas of investigation: starch breakdown and protein breakdown. We must now try *to explain the sorted observations*. For example, we could write:

'The contents of a cell contain an enzyme which breaks down starch.'

This is the *hypothesis*. It tries to explain what has been observed.

The hypothesis is then *tested by experiment*. Notice that this hypothesis confines the experiment to investigating the contents of a cell and its action on a carbohydrate. (The actions of saliva and of stomach and pancreatic juice could be the basis of other hypotheses and experiments.)

Sometimes the hypothesis to be tested is written in a longer form which states the results of a previous experiment. For example, we might write:

'*Hypothesis to be tested*: We have discovered that boiling destroys enzymes. Boiling the contents of a cell will affect its ability to break down carbohydrates.'

In summary, then, a hypothesis is *a statement based on observations that is tested by experiment*.

The aim of an experiment

Sometimes investigators design an experiment around a specific aim rather than to test a hypothesis. For example, one of the experiments in Chapter 2 of this book is headed:

'*Aim*: To investigate osmosis in eggs'.

Like a hypothesis, the aim defines the experiment.

Making a table for results

Recording results in a table is useful because:

- recording can be done quickly
- patterns may be seen more clearly
- the presentation of results in graph form or any other analysis (such as feeding them into a computer) can be done quickly.

When making a table, stick to the following rules.

1 *Keep it simple*.
2 *Identify what your experimental conditions are*. Make lists of these, representing the conditions by a few words. For example, in an experiment where there arc three tubes incubated at different temperatures the lists would be:

Tube	Incubation temperature/°C
A	17
B	37
C	80

Note that units of measurement (for temperature, in this example) are written in the column heading.

3 *Identify what you are going to record* – for example, the time taken for an enzyme reaction (such as milk clotting) to have taken place. This is the next column or columns:

Tube	Incubation temperature/°C	Partial setting time/min	Complete setting time/min
A	17		
B	37		
C	80		

Remember that you need to draw up your results table *before* you start taking observations.

Plotting line graphs

Look at the map of part of an American city in Fig. A1. How would you describe to another person how to get from point X to point Z?

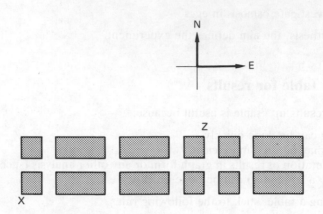

Figure A1

You might have said 'go four blocks east by two blocks north'. We use a similar strategy to fix a point on a graph.

You always start by drawing two straight lines perpendicular to each other. These are the *axes*. Look at Fig. A2. The point where they meet is the *origin*.

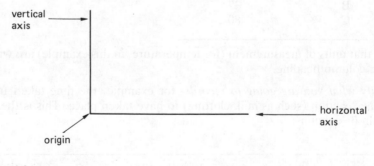

Figure A2

The axes are drawn on squared paper called graph paper.

The length of the axes depends on the highest and lowest figures you have to plot. This will also determine the scale of your graph. Let's consider a set of results from an experiment designed to investigate how the rate of a reaction changes with temperature.

Rate (in arbitrary units)	Temperature/°C
0	0
1	10
2	20
4	30
8	40

Notice that the highest and lowest temperatures are 0 and 40°C. The horizontal axis must be able to cater for this. If we use a scale of 10 °C = 2 cm, the lowest reaction rate is recorded as zero arbitrary units and the highest rate at 8 units. The vertical axis starts at 0 and will go up to 8; every cm is equivalent to one of the units. Look at the axes in Fig. A3. Copy the axes on to a piece of graph paper.

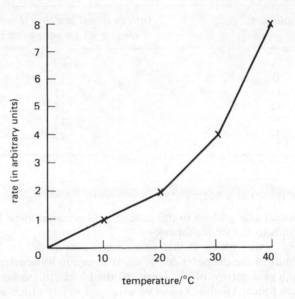

Figure A3

Now we can plot the points. We have recorded a reaction rate of 1 unit at 10 °C. To plot the first point, go along the horizontal axis to 10 °C. Move up from this point until you reach a point level with 1 unit of rate on the vertical axis. Put an encircled small dot ⊙ or a cross ✗ to mark the point. Look at Fig. A3, and find the first point.

Other points are plotted using the same principle. Look again at Fig. A3 and pencil in the position of each point on your own graph – you could draw lines from the points to the axes.

When all the points have been plotted, join them up using a sharp pencil.

Constructing bar charts and histograms

Bar charts

When you construct a line graph the variables both have numerical values, such as temperature and reaction rate. Sometimes, however, you may have results where one of the variables is not numerical – for example, if you measured the percentage of vitamin C in several different kinds of fruit.

Let's consider an experiment investigating the antibacterial activity of six toothpastes. Equal amounts of the toothpaste are placed in small wells in an agar plate inoculated with bacteria. If the bacteria do not grow around a well, this is an indication of that toothpaste's antimicrobial activity. The results might look like this:

Toothpaste	Antimicrobial activity: diameter of clear area on agar plate/mm
A	20
B	18
C	15
D	14
E	13
F	11

To present these results as a bar chart:

1 Draw a horizontal axis – which in this case should accommodate 12 intervals. This corresponds to the six toothpastes.
2 Draw a vertical axis – which in this case should accommodate 20 intervals. This corresponds to the diameter of the clear areas, in millimetres.
3 Plot each result as a narrow block. Draw all the blocks the same width – but do not let them touch. The blocks can be arranged in any order. Look at Fig. A4.

Histograms

Sometimes you may have to handle results that look like the table opposite, which is a record of the heights of a group of schoolchildren.

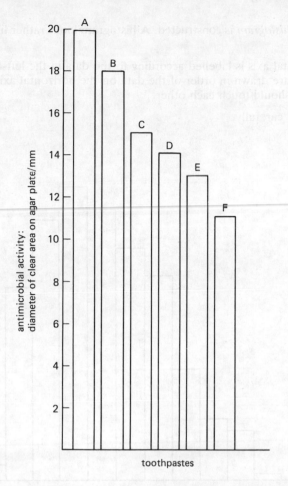

Figure A4   Bar chart representing the antimicrobial activity of six toothpastes

Height/cm	Numbers
120–125	1
125–130	9
130–135	11
135–140	20
140–145	37
145–150	42
150–155	49
155–160	53
160–165	47
170–175	35
175–180	3

In this case a *histogram* is constructed. A histogram looks rather like a bar chart, *except*:

- the horizontal axis is labelled according to the data in the left-hand column
- the blocks are drawn in order of the data on the horizontal axis
- the blocks should touch each other.

Study Fig. A5 carefully.

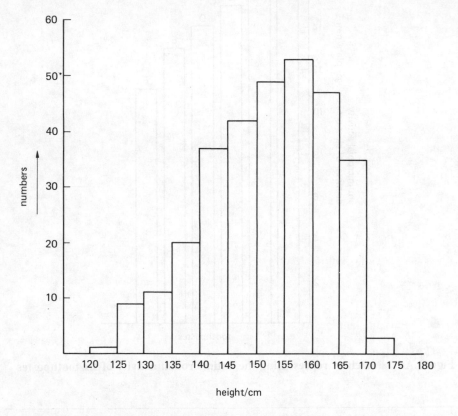

Figure A5 Histogram representing height in a group of schoolchildren

Atoms, elements, molecules and compounds

Substances are made up of *atoms*. Atoms are very, very small particles which are not easily broken up.

An *element* is made up of atoms that are all the same. There are over 100 different elements. The atoms of each element are different from the atoms of any other element. For example:

- *Hydrogen* is an *element*. It is made up of hydrogen atoms. All hydrogen atoms behave in the same way.
- *Carbon* is an *element*. It is made up of carbon atoms. All carbon atoms behave in the same way.

Carbon atoms are *not* the same as hydrogen atoms. They are a different size, and they do not behave in the same way.

Some of the elements you will come across while studying human biology are carbon, hydrogen, oxygen, nitrogen, calcium and sulphur.

Most atoms join together to make *molecules*. In the molecule of an element, all the atoms are the same. If the atoms in a molecule of a substance are of different kinds, the substance is called a *compound*. Look at Fig. A6.

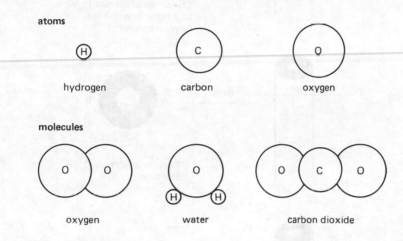

Figure A6   Atoms and molecules

Notice that:

- Oxygen is an element. An oxygen molecule is made up of two atoms of oxygen.
- Water is a compound. Its molecule is made up of two hydrogen atoms and one oxygen atom.
- Carbon dioxide is a compound. Its molecule is made up of one atom of carbon and two atoms of oxygen.

Transverse and longitudinal sections

A thin slice of a piece of tissue to be viewed under a microscope is called a *section*.

A section can be taken in several ways. For example, a blood vessel could be sectioned in the ways shown in Fig. A7.

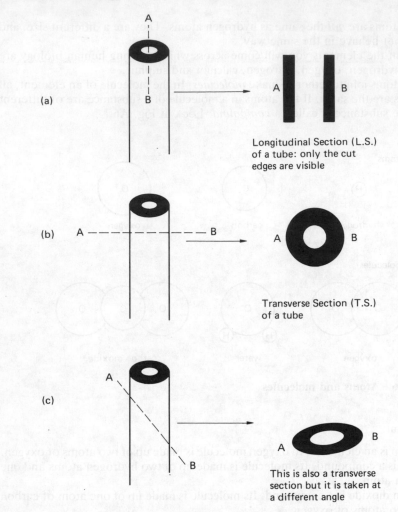

(a)

Longitudinal Section (L.S.)
of a tube: only the cut
edges are visible

(b)

Transverse Section (T.S.)
of a tube

(c)

This is also a transverse
section but it is taken at
a different angle

Figure A7   Longitudinal and transverse sections

Try making an L.S. and a T.S. of a carrot. They will probably not be thin enough
to view under a microscope, but cutting them will help you to remember what the
terms *transverse* and *longitudinal* mean.

Index